I am transformed and healed as I recognize my connection with the divine, my protector, archangel Rafael.

Advance Praise

Cancer is ubiquitous in every society affecting every family in every country. Tremendous strides have been made in our understanding of the treatment of cancer. New mechanisms of treatments have been approved with immunology, vaccines, targeted therapies being developed. This has led to an increase in survival and quality of life. Despite this, we still have big gaps in our understanding and little improvements in cancer treatment for some cancers. Dr Nezami presents his understanding and new interpretation of the current state of cancer treatment and its shortfalls. This book will validate some people's opinions and be controversial to other people. Until we can cure and control all cancers, there is room for improvement and new treatments.

—Steven Hager, D.O.
Medical oncologist and director of research at cCare (California Cancer Associates for Research and Excellence)

TRANSITION FROM *CRIME* TO *CARE* IN ONCOLOGY

A Critical Review of Current Cancer Research and Its Applications

MOHAMMAD A. NEZAMI, MD

CANCER FACTS PRESS

Sponsored by *Cancer Research and Life Foundation*
CRLFoundation.org

Published by Cancer Facts Press
Sponsored by *Cancer Research and Life Foundation* CRLFoundation.org

CANCER FACTS PRESS

Sponsored by *Cancer Research and Life Foundation*
CRLFoundation.org

Printed in the United States of America
10 9 8 7 6 5 4 3 2 1

Produced by GMK Writing and Editing, Inc.
Managing Editor: Katie Benoit
Copyedited by Liz Crooks
Proofread by TK
Text design by Libby Kingsbury
Cover design by Libby Kingsbury
Printed by IngramSpark

Print ISBN: 979-8-9861737-0-2
Ebook EISN: 979-8-9861737-1-9

To my current patients…

My future colleagues with “outside the box” thinking…

The 13,146 oncologists currently practicing in the US, who are the subjects of this book…

Future colleagues…why? In the past and present, only a few pioneers in my profession have been able to fully understand me. Practitioners who only practice medicine and refrain from the application of research and up-to-date information about cancer biology in their practice have a hard time understanding customized oncology, let alone applying it.

Customized programs targeting the real-time genetic character of a tumor appeal to those people who are elusive and defiant—the risk-takers and disruptors who break the rules and laugh in the face of convention.

When the established rules become a constraint to creativity, it takes individuals of vision to carry the game forward.

Contents

Author's Note

The intention of this publication is not to replace the formal training required for oncologists so that they may understand my therapy. Therefore, I have only introduced concepts to trigger the interest of open-minded providers who may then further expand their knowledge by studying my more than 100 publications. In the end, I hope the reader appreciates the fact that the intention of this manual is to engage him in a reality check—to enhance the concepts and disclose the misconceptions in oncology care as it is practiced today—as well as to further promote patients' public education so that they can be their own advocates.

Introduction

Right now I'm sitting on a bench in my backyard, and I'm remembering everything that has happened in my life. I'm thinking about how I became who I am today.... I think everyone has a story that is interesting and exciting, especially people who have gone through suffering and, at the end of the journey, have found happiness. We all pay a price for our success. No matter what we do, professionally in life, or how we live our life personally, there are always points in life where we need to *choose* between our happiness and the happiness of others. My choice has always been others. I derive a sense of gratitude and pleasure from seeing others happy.

It appears to me that happiness is the fruit of persistence and sacrifice.

All my life I hoped that one day I would achieve a humanitarian goal, to be able to be proud of myself and never imagined that I could accomplish what I have because what I think I have found is beyond all my wishes and as I'm trying to find how I deserved to be so proud of what I have done I cannot find an answer. Was it my difficult childhood? Was it the difficulties that I had when I lived back home? Or was it the difficulties that I had with the educational part of my life? I really don't know what made me special.

At one point I was revealed secrets from nature. Secrets that nature does not reveal easily, and if it does, it does it to the chosen ones. For the people who have not been chosen, it is all about mystery. When I heard about prophets and scientists in the world who became famous I always thought it's about luck, persistence, and education, but now I know it's beyond all of those. It's not about what you do, it's about what nature decides to do and the decision of nature is not something you can logically count on. It will happen one day and you really won't know how it happened, but that was what happened to me. Nature revealed its secrets to me, to let me know how I could find the answer to a devastating challenge for human beings: cancer. I was chosen to be revealed what weakness this giant monster has. I was chosen to be revealed to how to use that weakness to overcome the beast and ultimately help people.

The belief that life is a mission and the journey is the destiny was materializing in my profession, on every day of my working life. So what made me interested in treating cancer was the fact that I was always looking for an answer for the most complicated disease. By definition and character cancer is very devastating, as it usually affects otherwise healthy people and is detrimental, as most patients face a significant change in their life. As they go through the treatments they become more devastated because most of the conventional treatments have severe side effects and they take away the quality of life and unfortunately even with that, they are rarely effective.

My interest in integrative medicine mainly originated from my prior work in the state of California corrections system. Incarceration, with its level of stress, dramatically impacts the immune system. Generally speaking, incarcerated people are immune suppressed and chronic conditions such as cancer are very common. In the prison that I used to work in the rate of chronic conditions was beyond imagination, and in many cases I treated patients with cancer that basically originated from the level of stress that they had. Stress is a big immune suppressant and the quality of life and the diet experienced by an incarcerated person doesn't help either.

On my first day of treating patients in prison I noticed that my prisoners were patients with multiple complaints related to one disease. These patients were stressed, they were depressed, and many of them had had undiagnosed endocrine disorders for years.

A woman who had cold extremities, dry skin, hair loss, constipation, slow heart rate, and depression was seen by 10 different specialists. Ten different doctors were in charge, one for each symptom. A dermatologist was addressing her dry skin and hair loss, a psychologist was addressing her depression, a neurologist was looking for answers for her neuropathy, and a cardiologist was looking at the slow heart rate. A gastroenterologist was doing colonoscopies for constipation, on a patient who was over 75 years old (meaning a life expectancy of less than 10 years). Unnecessary procedures, and multiple doctors with different specialties. What I found was that all of this woman's symptoms were being caused by hypothyroidism, which was the result of premature aging and stress, which deteriorate the thyroid function. Instead of seeing all of these different doctors, the woman needed just one provider to connect the dots. This is how I could also relate the immune suppression with endocrine dysfunction and cancer: starting with looking at the patient as a whole and thinking about the cause of the problem rather the symptom.

So my intention in becoming involved with treating cancer patients was the fact that cancer was (and is) the most challenging disease of the time and the conventional treatments were the least effective. The other reason I was interested in treating cancer was that I was always searching for how we can treat the cause of a disease, instead of treating the symptoms. Conventional therapies generally address the symptoms, not the cause.

In cancer the cause is usually ignored. As a matter of fact, all of the conventional cancer treatments are intended to treat the tumor/cancer cells and none of them target the cancer mother cells or stem cells.

I have always lectured other physicians to focus their treatments on what originally made the cancer cells grow (what we call in medicine carcinogenesis), and I think I was the first one who discovered the treatment for carcinogenesis, or an actual way of reversing the DNA damage that caused the cancer from beginning.

So that starts from the time that I discovered a formulation—a formulation that was basically unknown or undiscovered up until that point. It appears that most of the scientists in the field of cancer have tried for many years to find the treatment for cancer but they have looked into the areas of interest that were determined largely by big pharma, companies producing new medicines, many only looking for big profits and not real solutions.

As much as these companies now are trying to make new chemicals, decades ago they initially tried to make new chemotherapies with herbs and drugs. They tried to extract the chemicals from the herbs. The idea was to treat cancer with chemicals from natural substances, and make huge amounts of money by selling these new chemicals that supposedly cure cancer. There are only a few types of chemicals that work naturally at all, and of those, only three or four types of cancer respond to the available chemotherapy and go into complete remission, what is technically used for the term "cure" in medical terminology, and even for those types (testicular cancer, for example), many patients still resist and recur.

The corruption is not limited to drug companies (I will further elaborate on this subject later in this book). It actually infects academic institutions as well, such as the University of Texas MD Anderson Cancer Center (MDA) and beyond. I personally had a very interesting report with them, when they illegally froze a research contract and fired their own scientist who was working with me. Since the subject of this book is not aimed at whistleblowing the establishment, I refrain from discussing the case here, but I refer to other cases reflecting the conflicts of interest.

Let's start with misinformation: MDA had claimed that they had a better than 50 percent cure rate (Annual Fund letter 1998, Politics of Cancer, revisited)[1] against Title 18 of the United States Section 1341 to fund solicitation and was distributed through the US postal services (Henry Novac letter, Politics of Cancer, revisited).[2] That was when Congress had designated the Practice Outcomes Monitoring and Evaluation System to overcome the threat posed by alternative therapies to the cancer industry. That said, MDA used the 11th Amendment to try to protect themselves from lawsuits from citizens of the same state or other states. (This was also the case when MDA refused to respond to my own inquiries about canceling the research contract I had with them.)

One would think agencies like the FDA and academic institutions are supposed to be looking out for the best interests of people. As such, my experience with the FDA through a workshop I attended in 2012 was very telling. As the speaker from the FDA, who was in a military-like uniform, was proudly lecturing about how FDA is responsible for making sure drugs coming into the market are safe and effective, I raised my hand and questioned why the last drug that had been approved for colon CA (Stivarga) has 40 percent toxicity and 0.4 percent efficacy (defined as an additional nine days of living!). The speaker suddenly admitted: "I have been telling my staff to avoid approving these '*toxic placebos*.'" That said, the drug is approved (in expedited pathways) and yet the FDA is blaming "alternative therapies" as ineffective!

The FDA's take on accelerated approvals these days is becoming a more and more dangerous path. For example, despite the lack of benefit shown in the IMpassion131 trial on using azetoluzimab, the drug was approved conditionally, along with paclitaxel or Abraxane, for use in triple negative breast CA, until a recent public push and whistleblowing convinced (or forced) pharma to withdraw it from the breast cancer market. The accelerated approvals for checkpoint inhibitors turned out to be a complete disaster after many of the trials contingent upon which the FDA had approved the therapy were halted due to safety concerns or inefficacy. At this point the FDA could *not* cancel the approval; only the drug company could voluntarily pull the drug, and they generally do not do that.[3] As an industry, oncology comprises 40 percent of all drug company activities in the US!

Based on NCI estimates, the environment plays a role in cancer incidence for 86 percent of cancers. Other factors are diet (35 percent); smoking (30 percent); sexual behavior (7 percent); occupation (4 percent); geography, alcohol, pollution, and chemicals (2–3 percent each); and medications (only 1 percent). This is in fact when we see in the news every day that medications widely used in clinics cause cancer. These range from hypertensive medications, such as ARB inhibitors, to such a simple drug as metformin, and even Zantac. Or we hear about radiation to the neck to treat thyroid lymphoma causing thyroid cancer, or radioactive iodine to treat thyroid cancer-causing secondary malignancies.

Now, with all the harm we know that common medications cause, and the fact that most cancers are caused by medications prescribed for other conditions, and beyond that when you read articles in the so-called ASCO Post discussing the subject of end-of-life care, encouraging a green light for medically assisted death, where a physician (an expert in killing), can assist suicide;[4] yet the same patient faces resistance when she desires to try a potentially life-saving (but not FDA approved) natural therapy! That said, we have trials using natural compounds with the exact same idea.[5] Even some cancer centers, such as the Memorial Sloan Kettering Cancer Center in New York, are using integrative therapies as part of a multidisciplinary approach to treat advanced cancers, such as colorectal CA, with the rationale that these therapies and services "were found to reduce side effects of traditional therapies, in some patients with CRC."[6]

How can someone support assisted suicide but resist any alternative therapies a patient desires to try?? No wonder there is such a conflict of interest in works published by MSKCC and the cancer industry. In 1988 at least 20 members of MKSCC had strong bonds in pharma, including the directors of Merck, Pfizer, Squibb, Bristol-Myers, and so on.[7]

It is hard to believe that in 2021, most cancers are not treatable or curable, and most of the chemotherapies used cannot increase or prolong overall survival, especially in patients with advanced cases (mainly because they increase intratumoral hypoxia as well as switching the tumor's metabolism from glucose to lipids).

Still, unfortunately, they are used in a wide spectrum in the medical community with very low success. That fact by itself made my mission, back at the time I started my career, to find a solution, as there was no other challenge in the medical world that would be more rewarding if somebody could solve it.

I was looking to be rewarded by seeing my patients' health and joy. So I started to investigate (very similar to how an attorney does) to find clues: what initiates the cancer (what are called carcinogens), why we get cancer, and instead of treating what is already there, how we might treat the cause of the disease, so that the disease will vanish. With that concept (completely different than the usual concept in medicine), I started my research.

I started to self-train and participate in every type of oncology scientific conference I could find. Finally I became formally trained in integrative cancer therapies and was certified. I listened to every single lecture I could find and searched every single therapy that was available and talked to every scientist in this field that I knew of, from Harvard stem cell research, to MIT, to a French scientist working on epigenomic coding and the SU2C Epigenetics Dream Team members themselves, whom I met in 2012 in the Koch stem cell research department at MIT in Boston. I was quite pleased when I saw that the Epigenetics Dream Team (including Stephen Baylin from John Hopkins) are working in parallel with me, with the same concept and similar R & D.

I really wanted to know what has been missed and why all of medical society is challenged by this disease. After becoming an expert in the science of cancer, I started collectively gathering information about several nonconventional treatments that were rarely available in this country and trying to collectively use those treatments and find out if they had any additive and synergistic effect, by themselves or with traditional care.

It is just like when you have the flu or a cold and you taking a remedy or a medication for it and it doesn't work. So you take another one and that doesn't work. You may even take a third one that doesn't work. But if you combine the medications they work, and that's what happened in my case.

The very first patient I treated for cancer had terminal refractory stage IV disease, and every chemotherapy already had been tried and tested on her and she had failed. She was at the point where her body would not respond to any conventional treatments. She also had tried all of the so-called "alternative therapies" available as well, which she had also failed. Her only option was hospice. But another famous physician, who knew me and knew that I liked to treat challenging cases, referred her to my clinic.

This was in 2009. The patient was very young, but devastated and sick with liver failure due to her cancer. Fortunately, her family was very supportive. Her husband had a position in a large company and they could afford anything that needed to be done. But there was really nothing that could be offered, except my newly designed treatment, which at the time had never been explored. It was simply an accident. The accident happened when I was treating the cancer along with her liver problem, and the medications that we used interestingly worked together on the cancer. The combination of the medications used for her liver and her cancer became a new drug, a new treatment, and a new way of treating cancer.

The patient's response to the therapy we tried on her was dramatic. She clinically improved after just a few weeks. We still had things that we didn't know about the therapy (especially the mechanistics), but clinically there was so much to see that one could not ignore. Her ascites and jaundice resolved very quickly and she became a new person. Her liver enzymes came down and her quality of life improved shortly thereafter. Her cancer had responded to this new therapy.

Now when I look back I know how and why my treatment worked, when at the time I really didn't know. It is just like any other discovery. When you discover something you have no information about the mechanism of action. Retrospectively I know how it worked, but it took me five years (of experimental therapeutic research) to understand what happened that very first day.

One thing I know for sure is that every patient who comes to see me is led to me by divine intervention. I personally do not believe there are any accidents in life, and in fact when I look back on my professional life and discoveries I am very much reminded of the movie *Awakenings*. The history is made by people, and who can tell the story better than the patient herself? This is why we hope to make a documentary to best tell the story. Here are some cases we will be looking at:

1) A 24-year-old woman who was diagnosed with schizophrenia at the

age of 18 and had been treated since then with heavy doses of antipsychotics with no results. She was suffering so much that she was suicidal. Then someone referred her to us. I questioned the initial diagnosis and ordered a brain MRI and discovered that she had a large brain tumor. I referred her to two surgeons (the best at Stanford and University of California, Irvine) and both said they could not remove the tumor as it was too big and in a critical area of the brain. She would die if they attempted to resect it. I started her on my discovery of MTET, and in a matter of 2–3 weeks her symptoms started to wane and we were able to videotape her progress. We stopped all her medications in a matter of a month or so, and now as we speak, she is symptom-free (cured schizophrenia!). This is a perfect example of a case that demonstrates how the medical system only focuses on one thing, and that is the symptoms of diseases, rather than their cause.

2) A 17-year-old girl who was diagnosed with sarcoma. This girl was referred to me in 2014 after she had tried all other modalities of care including chemotherapy (more than 16 rounds), radiation, and several surgeries. She was seen at the best institutions, which could only offer her more suffering. She even went to Germany and was treated with immune therapies, with no results; she just got worse. It was at the German clinic that someone told her parents about me. Interestingly, she lived only few miles from my clinic but yet did not know about us. Immediately her parents brought her back to California to see me, but it was too late. She was dying. With all that I did, I could not save her. She died after about 1 year under my care. Although her family was very appreciative of me for keeping her alive as long as I did, this was not enough for me. I struggled with why God should send a patient to me when I could not save her. I attended her funeral and up to this point I had never seen such an impactful life and impactful death of a patient. Less than a year after the girl passed away, her mother came to see me for an enlarged lymph node in her neck. I could not imagine that this could be malignant. For 2 weeks she was treated with antibiotics and finally we decided to order a biopsy to "rule out" cancer. Well, the biopsy came back showing very aggressive (double hit) lymphoma. She started our therapy (MTET) immediately and since then (2016) she has been under my care. Technically speaking she had less than 6 months to live at the time of diagnosis, and now it has been over 6 years and she is doing great. She

is teaching volleyball and her cancer is completely controlled. She is now 51 and she will never die of cancer. To go back to my question of why her daughter was sent to me when I could not help much, now I know; it was a sacrifice the mother had to make in order to learn about me so that I could save her. I call this reverse destiny, as usually children learn from their parents' mistakes, and not the other way around. I wish her daughter had come to me sooner instead of trying conventional therapies, so I could have saved her too.

3) A 19-year-old boy with advanced cancer of the bone (osteosarcoma). This teen was treated for 2 years at Children's Hospital of Orange County. (As we know, osteosarcomas peak in young adults and also in elderly people in association with Paget's disease, and in both groups the long-term survival rate is less than 20 percent and the role of chemotherapy post-surgery is questionable. The agents commonly used are ifosfamide and etoposide.) In this case, more than 90 tumors were seen in his lungs after he had exhausted surgery and chemotherapy, and he was devastated. A very young football player who otherwise could have a successful life and career, he had become disabled and terminal after doctors had tried every chemotherapy under the sun on him. I got involved in his care after he was referred to me by an oncologist who had followed my work for years, although I did not know her. She called me and asked me to save this child and I accepted. He was transferred from the ICU to my clinic and on the very first day I learned that he was extremely sick, not having being able to eat for weeks, and was unable to walk or even breathe. Immediately I started him on the therapy and immediately he responded. In a matter of a few days, he was able to be off oxygen and was able to walk and eat. We called a movie crew in Hollywood that I had met and they came by and started filming him. It was an incredible story (like *Awakenings*) until we found out that the boy had been admitted to the hospital for a possible infection and his doctor there refused to allow him to continue to receive our therapy. I will never forget his last hug, when he said to me I was the only one who cared. I visited him in the hospital and saw a 19-year-old gasping for air, while his oncologist was denying him simple oxygen. I will never forget his funeral, when his mother, who happened to sing in the church as her profession, sang with her son's body lying in front of her. It was such an amazing and unbelievable

moment of the harsh reality of medical barbarism defeating the basic right of humanity: *life*.

Let's go back to my journey. Through the years, I had found ways to optimize the treatment and customize it. As I have described, communicating with different scientists in the field helped me a lot. The protocol that I discovered was a combination of several different modalities and methods of treating cancer but yet was completely nontoxic. I gained a new world of confidence when I investigated the biological application and understand the mechanistics of the therapy. This is quite in contrast with the concept of "combination" therapies used in conventional settings, where there is almost never a survival advantage, and almost always additive toxicity.

Even in the most established protocols that combine chemotherapy with radiation, we now know that each modality makes the tumor resistant to the other; for example, radiation makes a glioblastoma more resistant to chemotherapy and vice versa, so much so that in a recent study on anaplastic gliomas (1p/19q), the use of Temodar alone had a small advantage when used after radiation (adjuvant setting) and not concomitant with radiation.[8] That said, the Stupp protocol still is the standard of care for gliomas, combining chemo and radiation.

I will be discussing the traditional modalities of care in this book in detail and will try to open your eyes as much as possible about both conventional care and care that is even worse, as is the case for the more recent therapies. Undergoing chemotherapy is dying a slow death, in my opinion. We now are studying many types of cancer and their mechanism of resistance to therapies and we have realized it is not the tumor itself, but rather the therapies used, that make the disease more resistant. For example, we now know that TP53 mutated clones emerge after a patient receives chemotherapy, triggered by DNA damage causing tumor clonal expansion with higher resistance. Recently *Oncology Times* published an article about how standard treatments fail,[9] explaining the mechanism behind the cancer's resistance when exposed to cytotoxic drugs by the extracellular matrix and enhancement of fibrosis in the tumor after treatment with chemotherapy. I have also published articles in this area looking at different mechanisms of action, by increased intratumoral hypoxia, after using chemotherapy. (Interestingly, articles in this area were somehow eliminated from the internet, including a study looking at the same phenomenon after the application of Sutent and another similar study after exposure to Cytoxan.) Epigenetic therapies, on the other hand, reduce intratumoral hypoxia. As such, I propose using 18-FMISO scans, which can identify intratumoral hypoxia and research suggests that if a tumor responds to treatment of any kind with reduction of the PET FMISO, the treatment strategy of radiation and chemotherapy can be de-escalated (or omitted). This was recently published in *OncologyLive*.[10]

We are also looking into using chemotherapies less and less in clinical practice, starting with breast cancer. Even in the most obvious cases, where we thought for the longest time that the tumor response was mainly due to the combination of targeted drugs and chemotherapy, we are now questioning the entire concept.

For example, regarding HER2+ breast cancer, a very large study by José Manual Pérez-García, published in the *Lancet* in June 2021,[11] showed that de-escalation therapy, with no use of chemotherapy at all, had the same results as targeted therapies to inhibit HER2 alone, confirmed by the pathological response rate and FDG avid PET scans, suggesting that early stage HER2 breast cancer is likely to benefit from chemotherapy-free targeted inhibition of HER2.

This is also true about other types of breast cancer. In fact, we now see study after study that suggests that the positive effects of chemotherapy on some patients is *not* related to its effect on tumor rater, a type of collateral damage that happens in the ovaries (ovarian failure due to chemotherapy), which automatically suppresses the female hormones feeding the cancer. This explains why patients at times survive the disease and chemotherapy gets the artificial credit, where in fact a simple pill to suppress the hormones could have done the same.

And finally, in patients who are taking aromatase inhibitors, studies now have shown that using these drugs for an additional 2–5 years did not improve survival beyond the initial 5-year limit.[12]

Chapter One

To Recently Diagnosed Patients (Disease and Destiny)

As a physician, every time I address a situation where there is bad news for the patient I try to place the context and the relativeness of the news into the broader picture. I want to start this chapter by talking a little bit more about the biology of the brain before I start to talk about the initial shocked reaction that patients naturally have when they receive a diagnosis of cancer.

In every patient, when they are facing bad news (or simply looking at an unknown future), there are different parts of the brain reacting. This is a very important element of decision-making. A normal human brain is simply categorized into three levels of development. The first level is the brain stem, which is in charge of reactions and reflexes and is the most primitive area of the brain; for example, when you are in a position of extreme stress, such as an earthquake or an immediate danger in your environment, you react to the situation with a fight-or-flight reaction. In this situation cortisol is produced by the adrenal gland and the sympathetic drive overrides any process of thinking, and the entire brain goes into survival mode.

The second layer of the brain, the animal cortex or paleocortex, is on a higher level and is in charge of formatting one's behavior based on animal instinct. This part of the brain is in charge of and involved in many of the human reactions to an environment that is not necessarily dangerous but rather requires an animal instinct for decision-making. These might include protecting one's children, competing for survival through reproduction, or competing for survival when resources or food are limited in the environment. Unfortunately, because the evolution between primates and animals and later humans happened very quickly, humans still react to our environment mainly based on our animal instinct. As we discuss this further you will see that in the higher level of the neocortex (where most sophisticated decision-making is formatted) logic and reasoning start to become involved, but the paleocortex still plays a major role, as the majority of human decision-making is aimed at survival.

The other subject of the discussion that is particularly important is related to the hijacking of the neocortex, which is involved with decision-making, by the paleocortex, which is the animal brain. Many criminals who have acted based on intelligent decision-making and premeditation have used their neocortex to come up with the most sophisticated forms of destruction based on their animal brain controlling their sophisticated human brain. This process of hijacking, when it happens, makes humans the most dangerous species.

The bottom line is that although we as human beings always differentiate ourselves from animals at large due to our brain function, we should admit that most of our decisions are based on animal instinct and not cortical thinking. This is scientifically in concert with having 93 percent of our DNA copied from animals, and only 3–7 percent so-called "human" DNA.

Regardless of what creature we are discussing, from plants to animals to of course humans, a general concept regarding the process involved with reacting to a stimulus (in this case the bad news of a cancer diagnosis) involves a series of steps: Reception, Processing (Thinking), Decision-making, and Action.

For example, after a plant is exposed to a new environment, such as sunlight, darkness, or oxygen, it receives the stimulus, then processes it and forms a decision before it reacts. This process is very well-known in the most primitive creatures, like plants. Yet we see in the most sophisticated creature (humans) that this process is cut short, meaning that depending on the stimulus of the environment the decision-making and thinking processes quickly disappear and the stimulus can cause a reaction without those processes happening. For instance, the human may reflect on their diagnosis of cancer without being able to analyze the information accurately.

The reason behind this expedited reaction is the fact that the human cortex, instead of slowing the reaction and processing it quickly, defines it as a fight-or-flight situation. Any time this happens, the human responds without thinking. Unfortunately, one major difference between humans and animals is the collective anxiety that is seen in humans more widely than in any animal.

I had to elaborate on this subject, as I feel it is very important that newly diagnosed patients comprehend how important it is to understand that the *fear* related to the news of their cancer is naturally something that will automatically take away their healthy way of thinking and ability to assess the problem and resolve it. At times, it is also essential that the patient seeks supportive groups to help him or her control the fear and avoid making uneducated decisions based on it. I always encourage patients to seek support from their role models in their belief system. If they are religious, they should seek guidance from the establishment and know there is only one way out of this: faith. Faith and fear are enemies. I also believe that, as the Bible says, knowledge is power, and knowing that cancer takes away your control, you can have peace of mind when you know your provider is in charge and cares about you.

As the physician, it is the responsibility of the provider to explain this process to the patient and to make sure they understand why it is important for them to have self-awareness about their reactions. I'll give you an example: A patient is diagnosed with cancer in a routine screening, without having a background history. Up until that point, the patient lived a normal life, never even thinking that this could happen to her. As a stressful stimulus this news is quickly received, and without ensuring there is enough analysis made; the patient reacts by surrendering and starts treatment that she has not quite analyzed or thought through.

Unfortunately, when this happens—when the human surrenders without enough thinking—it is very important in forming the patient's destiny, because every decision that a patient makes from the beginning significantly impacts their outcome and their response to treatments in the future. The first step is very critical and vital to this process…and you *cannot* do it by yourself.

Unfortunately, in the field of cancer the more you are confident about your way of handling the stress, the less you may be able to relieve the stress; and the reason is that there are so many unknowns in the field of cancer, about both the whys and the hows, as well as the ifs and the buts, so that even the most intelligent patient at one point realizes that he or she is not able to understand the whole picture.

When I, as a physician, see a patient, I always encourage them to look at their destiny more carefully (regardless of their diagnosis of cancer); for example, my discussions with the patient at the time they are diagnosed with cancer or as they go through treatment are not only about the cancer itself, but also about what is it that they want to accomplish in life regardless of the diagnosis of cancer. This is something that patients with cancer always miss; as soon as they are diagnosed the only thing that they think about is their diagnosis and how they should fight the disease rather than knowing that the disease itself is not their destiny; rather, it is their destiny that defines their reaction to the disease and its diagnosis.

When I was in school as a child, we were told a good story about how we each look differently at life. On a winter day at school, as a group of children were playing in the playground, the teacher asked them if they could walk on the snowfield in a straight line from one side of the yard to the other and reach a tree on the other side with their feet making a straight line to win the game. Every student tried, but there was only one girl who could make a straight line, as every other person had made zigzag lines. When the teacher asked the girl what her trick was to making a straight line, she responded that every other student had looked at their feet while they walked and tried to make their feet straight, but she instead looked at the tree (her destination), and that was how she was able to make a straight line.

The same concept is true here; when a patient is diagnosed with cancer, they have to look at their destiny as if they do not have cancer; that's why I classify patients into two groups: the first group looks at the destiny of their life on a daily basis, without being distracted by the daily matters of life; and the second group only looks at the daily matters of life and tries to make their own destiny. The second group always fails to accomplish their goal.

When a patient is diagnosed with cancer, for their provider this is the most important element of their care: to try to coach them to look at their destiny and not be distracted by daily decision-making. In order to do that, they have to control their human brain, or neocortex, by understanding and accepting the fact that the majority of their decisions will not be made based on known factors; rather, they will be based on their surrender to the unknown factors.

You can see that there is a relief that comes with surrendering to what people call fate, as it significantly negates the effects of the fear of unknown. Accepting the unknown brings the patient peace, and so the provider's role is very important in bringing peace to the patient by doing two things: one, making sure that the patient looks at their destiny and not everyday happenings, and two, educating the patient about making informed decisions to the level of the known facts and letting go of the rest, and at the same time building the relationship to trust the provider. In other words, the patient has to recognize that they can only control what they can and what they know, and for the rest they have to accept the results with grace. I know this subject sounds a bit spiritual, but the reality is that for every spiritual matter there are chemical reactions that happens in the brain and specific molecules that impact the patient's well-being and physical destination. This part has been extensively explored in epigenetic modifications of human behavior and, as it relates to cancer, on how a patient perceives his or her disease and destiny.

It is our choice to bring the best destiny to ourselves by making informed decisions based on our limited information and to trust the rest to a higher power. Fortunately or unfortunately, with cancer the unknowns are much more than the knowns, even for the most sophisticated and intelligent physicians, and yet it is important to keep our focus on the tree at the other side of the yard and not our footsteps.

This process is best handled when the patient accepts the role of the physician as the leader and gracefully submits to the provider's leadership and accepts the outcome that it brings. It is as if the patient puts the physician in the pilot's seat and then relaxes with their seat belt fastened and goes through whatever turbulence happens, and accepts the fact that nobody can guarantee them a safe landing, and that has nothing to do with the diagnosis of cancer. Our life is a very short journey and I understand that at times it can be very painful to end, but it is our choice to make it a pleasant experience, with the best impact we can have on our surroundings, whether that is our family, friends, or even our physician, or make this the worst experience for them and still not be able to control the outcome.

In order to enhance the communication skills between the patient and the provider, I believe both have to be on the same page. One simple way to do that is to improve the sense of hopefulness for both when it comes to treating a patient with advanced cancer. I believe in Hippocrates' notion that at the very heart of the physician's technique lies the human dimension of his art, suddenly coming into view like a subliminal image that unexpectedly presents itself to one's consciousness. In everyday practice this dimension generally assumes three forms: gentleness in treatment, courtesy toward the patient, and conversation with the patient. After all, isn't that what medicine is all about?

Current studies show that a brief communication intervention that takes about eight minutes of a physician's time can impact a patient's survivorship.[13] Most importantly, physicians should participate in hope-developing workshops based on Snyder's model of hope, to enhance patients' desire to set goals, design pathways to achieve them, and follow those paths. These workshops are available to both patients and providers and can assess and in fact measure the level of hope through the Herth Hope Index and the Adult Hope Scale. This simple tool has increased patients' hopefulness and reinforced the desire to live in breast cancer patients.

Communication is also key when it comes to pediatric patients. There are several consequences of communication breakdown, including emotional distress, insufficient understanding, decreased trust or confidence, decreased hope and engagement, financial distress, and decreased access to resources. We promote six categories of support responses to these challenges, provided by the clinician: Exploring, Acknowledging, Informing, Adapting, Addressing and Advocating. In the absence of these supportive responses, parents may engage in confrontational advocacy, which can negatively impact the clinician–family relationship.

The pediatric population is a very sensitive area of medical oncology practice. Although the FDA Modernization Act of 1997 adopted a pediatric exclusivity provision approved by Congress, along with the Pediatric Research Equity Act (enacted in 2003), the recent Research to Accelerate Cures and Equity for Children Act (enacted in 2020) has made it difficult to process new targeted drugs in this population, requiring adult-like studies to get approval. There should be a better way to enforce enrollment of the pediatric population in clinical studies, which it is mainly omitted from at the time of this publication.

Chapter Two

Basic Practices of Oncology

In this chapter I would like to discuss some basic foundations for the best practices of oncology in the form of standards and beyond. As everyone knows, the standard of care is defined based on evidence and this evidence is defined based on knowledge extracted from trials (as the "practice" of medicine is defined by trial and error).

None of the proven facts in medicine were proven before being tested or tried; for example, nobody knew that using a specific medication could help a patient before it was tried, so the hypotheses are all based on "concepts," and the concepts are the simple mental process of a knowledgeable person who can make common sense out of different findings. As you can see, the most sophisticated technologies started with just a thought or an idea, but the most important thing is that this thought process (that creates the hypothesis; further hypothesis are tested on different levels and finally is tried on humans and later confirmed to be effective) is not a linear path; rather, it is a circle, meaning that it starts with the concept and then ends in the concept.

That is why most hypotheses only result in minimal discoveries and the majority of hypotheses generate further hypotheses, going on and on in a circle. Each time the hypothesis is regenerated only a fraction of it is proven to be effective or valid. The biology of cancer is very sophisticated and our understanding of it has just recently evolved.

We have realized that many of the concepts that we had in the past and based on which we treated patients were simply false. I always give the famous example of describing an elephant in a dark room, when a group of people who have never seen an elephant go into the room and touch different parts of the animal without seeing it, and then come out and describe what the elephant looks like. Obviously, some people are going to touch different areas and have different descriptions of what the elephant looks like. It does not mean that they are wrong, it means that there are many other parts they have not touched, and the same is true of cancer. Every scientist comes up with a theory, and it may be true in some areas and not true in many others.

The whole concept of treating cancer based on the available evidence lacks the outlook over the unknown evidence. Therefore, every individual who is involved with cancer care at any level must first admit that the unknowns are far more than the knowns; that there are lessons to be learned; and that the concepts they have based on the most validated evidence could simply be wrong.

When I applied this concept in my practice, I started with knowing nothing, but knowing nothing helped me to try to learn, versus the many oncologists who do not appreciate the unknowns. When they start seeing a patient, they already have a premeditated plan based on the things that they think they know, and that causes them to be blind about findings that they could learn on a daily basis.

Many of my patients have asked me where I learned all of this information, and how much of what I practice was taught to me at school. I always give them a simple and unique answer: What I learned from many years of practicing medicine and medical school is the same thing: what I am not supposed to do. Meaning that almost everything that I do in my practice is not what I have been taught, although I confess that the most important element of care should be based on the philosophy of do no harm. As everyone in the medical field in this country admits, the first cause out of three for patients' morbidity and mortality is medical mistakes. Worse than that, it is not medical mistakes that contribute to most patient morbidity and mortality. Rather, it is legitimate, correct medical care that causes morbidity and mortality. This is not just my opinion; it is actually published, and we all agree that it is true. With that said, I try to learn from each patient and absorb that information and apply it to the next patient. In fact, I have tried every treatment I prescribe to my patients on myself first.

I've had many patients in my life whom I treated but could not help, or I could not accomplish what I wanted to do many years ago. Yet all these patients shared one thing: they taught me something. Many of them, as I recall, agreed from the beginning that they would surrender. I will emphasize this word again and again in this chapter: *surrender*. Surrender to the outcome, understanding that the outcome is going to be either beneficial to them or beneficial to somebody else who will undergo my treatment in the future. I always encourage my friends to view the movie *Awakenings*, which was based on a true story and is extremely relevant to how I started to make a difference in my patients' lives. It shows how a drug that was experimental at the time was able to have such a huge effect based on just a concept, as well as the courage of the physician to try it.

Even if every door closes, the patients still believe in the outcome, and if the outcome is not in their best outlook they still are grateful for the efforts the provider has offered with his or her best intentions. This is very important, because it's not really about the outcome; rather, it is about the effort. If a provider puts in the effort, and the core of the effort is the patient's most desirable goal, in the end the patient appreciates it regardless of the outcome.

I would like to emphasize several lessons that I have learned from the patients in my clinical practice, starting from the first and most important concept of do no harm. As I explained and medical practitioners all agree, the first rule of medicine is do no harm. Unfortunately, some people in the medical field have provided so-called legitimate clinical care that causes significant morbidity and mortality in patients across the country, which is known to be the third leading cause of death in the US. We are not talking here about medical malpractice; we are talking about legitimate medical care.

It is hard to believe that practicing medicine as taught in the medical schools in this country has caused the third leading cause of death (using appropriately prescribed medications), and as such this country carries a higher number of medical errors than almost all other comparable countries in the world. If you wonder how much of what is practiced in oncology care falls into this category, all you need to do is look at the commonalities between the graphs on page 22.

This means that a major portion of the legitimate "evidence-based" medical care neglects the first rule of medicine. Obviously, all cytotoxic chemotherapy come with the burden of significant toxicity to the host. So again here, it is particularly important for a provider to balance the pros and cons of a treatment. Not only that, but also investigate what the core aim of

Cancer is among the three leading causes of disease burden in the U.S.

Age standardized disability adjusted life years (DALYs) rate per 100,000 population, both sexes, 2013

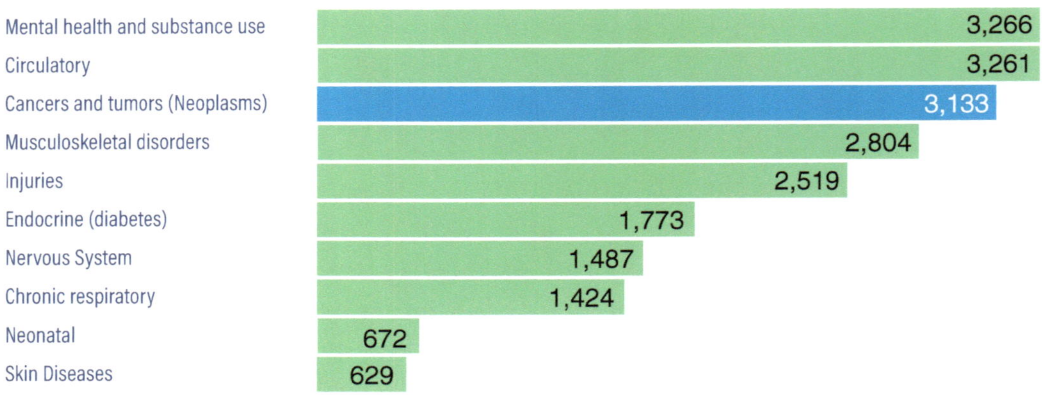

- Source: Kaiser Foundation analysis of data from Institue for Health Metrics and Evaluation, Global Burdern of Disease Study

The U.S. has higher rates of medical, medication, and lab errors than comparable countries

Percent of adults who have experienced medical, medication, or lab errors or delays in the past two years, 2016

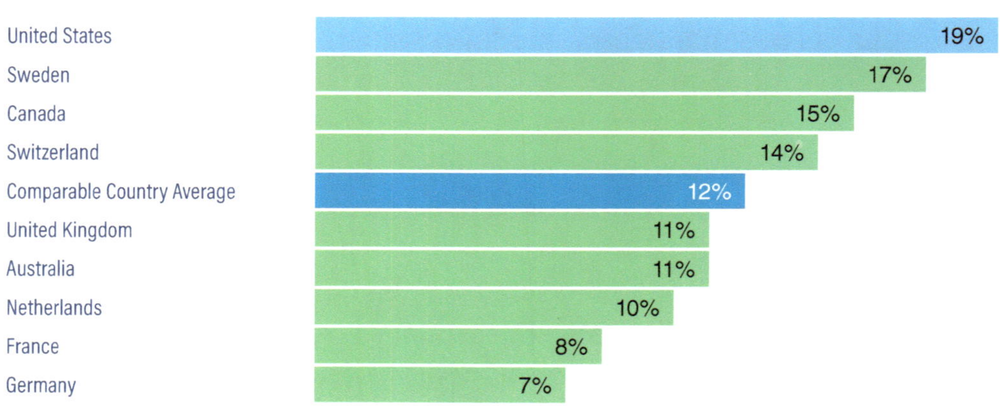

- Source: Kaiser Foundation analysis of data from Institute for Health Metrics and Evaluation, Global Burden of Disease Study

the treatment is in the eyes of the patient. In other words, when a pharmaceutical company comes up with a new drug for treatment of cancer, they only look at treatment of the tumor, not the treatment of disease in a person.

All cancer research coming up with new technologies and treatments looks at the tumor as if the tumor equals cancer. After 2005 and 2006, with the discovery of cancer stem cells, we all know and agree that the cancer and tumor are not the same. As such, treatments targeting tumors should be evaluated based on their effect on the cancer, as there is a dichotomy in the effectiveness of a so-called effective treatment for the tumor and its effects on cancer stem cells, which in my definition is the cancer.

I am now going back to the lessons that I have learned from my patients. Understanding the patient's aim of their treatment is important, as it may be quite different than the aim of the treatment by the physician. As surveys have shown us, for example in stage IV ovarian cancer that has failed to respond to the standard treatments, the quality of life is significantly more important than the quantity of life in the eyes of the patient. Although the FDA has approved drugs based on their positive effect on quality of life (e.g., Mitoxantrone for prostate cancer), this is only an exception to the rule. Patients' perspectives on their own care were explored in the book *Off Our Chests: A Candid Tour through the World of Cancer* by John Marshall and Liza Marshall.[14] The idea that *public* health pertains to the individual is disputed,[15] as there are critical differences between population and individual interest.

We have already discussed the fact that patients are looking for a better quality of life rather than an extended quantity of miserable life. We also argued against the common thought that cancer is a killer; in fact, the evidence suggests that treatments of cancer are the killers, as the more we treat patients successfully using toxic therapies, the more we impact the survivors' quality of life negatively. Currently, in 2021, we have over 17 million cancer survivors in the US alone. (We will discuss this subject in detail later in this book.) For example, women who are treated with chemotherapy for breast cancer have aged around 20 to 25 years by completion of therapy, if they survive. In other words, even if a treatment is considered effective by the best means of assessment (quote unquote "overall survival"), the treatment may fail to provide a better quality of life, and as such fails the aim of the patient-oriented practice goal. A perfect example is childhood cancer survivors who suffer from breast cancer in adulthood as a result of the mediastinal radiation used to treat their childhood cancer. Also, patients with testicular cancer can suffer from cardiovascular events (up to a 35 percent increase in incidence), over 10 years after their diagnosis and treatment with platinum-based agents (platinum levels can be detected 10 years after exposure). There is also significant linear risk of leukemia post exposure to these agents, associated with accelerated cellular senescence, reduced telomere length, epigenetic modifications, somatic mutations, and mitochondrial DNA damage.[16] Several tests, including the DunedinPACE test of aging, measure the allostatic load of lifetime exposures to an aging environment. Markers of senescent cells include increases in cell size, lipofuscin accumulation, high expression of the cell cycle regulators $p16^{INK4A}$ and $p21^{CIP1}$, and senescence-associated secretory phenotype factors (*e.g.*, IL-6, IL-8, monocyte chemoattractant protein-1, plasminogen-activated inhibitor-1, and many others), increased cellular senescence-associated β-galactosidase (SA-βgal) activity, and appearance of senescent-associated distension of satellites and telomere-associated DNA damage foci, among others.

Senescent cells do not divide and therefore cannot acquire advantageous mutations; as such, one mechanism of reversal for accelerated aging is using compounds that eliminate the aging cells. This strategy is possible by targeting the antiapoptosis pathways in aging cells.17 Senescent cell antiapoptotic pathways include PI3KCA, HIF, AR, ephrin receptors, MDM2, TP53, P21, Bcl-2, and HSP90.

This strategy is the mainstay of preventive oncology in my practice. A short exposure (a few hours) to epigenetic modifiers can maintain the antiaging effect for seven months. Interestingly, the epigenetic modifiers I use target most of the antiapoptotic targets, except ephrin receptors, which can be blocked by dasatinib. Bcl-2 inhibitors (like navitoclax) also have been suggested by James Kirkland and Tamara Tchkonia.[18] Another study by Omid Salaami at the Mayo Clinic showed daily dasatinib was more effective than weekly imatinib used in patients with cancer, to reduce their weight and hemoglobin A1c (as a desirable side effect).[19] The predicted weight loss (and senolytic/antiaging effect) caused by imatinib is related to its glycolytic activity. One of the mainstays of the efficacy of this drug is related to the metabolic effect, and I have published this information in the past, as well as on a wide range of other glycolytic drugs (2-deoxyglucose,

3-Bromopyruvate, etc.). Most oncologists, however, are unaware of this important clinical effect. Interestingly, using epigenetic therapies in combination with both imatinib and dasatinib causes durable responses in both TKI inhibition as well as glycolytic/senolytic effect (quercetin and fiestin, as a reference). The effect lasts for weeks until the aged cells are repopulated.

The other subject for female patients is the preservation of their fertility. Medications, compounds that reduce the cytotoxicity of chemotherapeutic agents, prevent stromal injury or reactive oxygen speciesROS, or block apoptosis generally improve follicular preservation. Curcumin, imatinib, GCSF, sphingolipids, dexrazoxane, shilajit, mangafodipir, mTOR inhibitors, and GSK-3 inhibitors (such as lithium) are good candidates in combination with proteasome inhibitors to preserve ovarian function and prevent premature ovarian failure for women trying to maintain their fertility. GSK inhibitors reduce the NF-kappaB expression and modify tumor progression. My young patients with cancer have the desire to become pregnant and preserve their fertility. As we all know, women at fertility ages have the desire to keep this ability despite a diagnosis of cancer. Many can—for example, in stage I ovarian cancer, the surgery can be modified to preserve the patient's fertility (by preserving the uterus and contralateral ovary), as studies have shown that complete surgical resection in stage I can at times increase the risk of recurrence.[20] That said, in stages above I, complete cytoreduction is suggested (less than 1 cm residual tumor), or for patients with breast cancer, the choice of chemotherapy can be selected based on this option. Recently we have confirmed through different studies that the addition of GnRH agonists to the chemotherapy for breast cancer is safe, and also can help to preserve ovarian function and fertility (as in the PROMISE-GIM6 trial).[21] This is when, for the longest time, the common concept was that a hormonal blockade could not be added to chemotherapies. Many studies had otherwise indicated that metronomic use of Cytoxan at 50 mg a day in addition to Femara (letrozole) as an aromatase inhibitor had a survival advantage in stage IV breast CA without significant toxicity. That said, the standard oncologist is still under the impression that if the cellular division rate is reduced by a cytostatic drug such as hormone blockers, the cytotoxic drugs are not as effective. This could be true when the cytotoxic drug efficacy is through a killing effect, but not in the case of metronomic use, where the effect is by the inhibition of angiogenesis.

As we learned more about the real benefit of using traditional therapies, we learned that many of the standard therapies are being overused and abused. The whole concept of maximum tolerated dose is now being questioned. Now we know, for example, that there is only a 4–17 percent at best improved 10-year survival rate when chemotherapy is used for breast cancer, in general, and in patients with low recurrence scores there not only is no benefit, the recurrence rate increases with the use of chemotherapies. In fact, the ASCO Post expert reviewer Charles Shapiro admits that "there is no surprise that patients who have received prior chemotherapies have lower survival" (by 0.6 months).[22] The same is also true for most of the so-called targeted drugs; for example, using bone-supporting therapies, such as zolendronate, clodronate, or ibandronate, or denosumab application post definitive therapy in breast CA, especially for premenopausal women, did not change the bone recurrence-free survival, or other end points such as time to first clinical fracture. Another example is the application of post-operative radiation to patients with stage III lung cancer after complete resection and adjuvant chemotherapy, which has resulted in no benefit in survival.[23] Recently Yukihide Kanemitsu et al. (2021) published data showing that addition of chemotherapy was beneficial (improving disease-free survival) to hepatectomy in metastatic colorectal cancer, when the data (in their own article) clearly states that the overall survival rate dropped from 83 percent to 71 percent when chemotherapy (mFOLFOX6) was added![24] Other studies have suggested drug holidays!

The whole concept of customized oncology or precision oncology comes with the notion that every individual—and for that matter, every individual tumor—must be treated differently. As such, the targeted gene testing is supposed to provide better quality-adjusted life years and increased cost-effectiveness ratio to QALY. On the other hand, cancer is the third leading cause of disability-adjusted life years.

As such, the information must be evolving, and not only that, but it also has to be dynamic to match the dynamic nature of the disease. The adaptability of a provider's decision-making is the most important element when you are treating a disease with such a dynamic characteristic.

Chapter Three

Handling a Difficult Patient

PRECISION (ULTRA-PERSONALIZED) MEDICINE, LIKE MANY CUSTOMIZED programs, is not designed for everyone or every mindset.

If I were going to list significant factors that I have learned from my patients, I would start with grace and gratitude. That said, this chapter will focus on practical information regarding handling the challenges in patient care, especially when the patient has a terminal diagnosis. I would like to emphasize the critical role of a provider in dealing with a patient with a terminal disease. Several requirements are primary. Number one, the physician must understand that listening to the patient is more important than listening to the patient's family members, as many times the family members have different expectations than the patient does. Also, it's important not to presume that the patient will have certain expectations, as every individual may be completely different in their stages of handling the disease and their goals of care. For example, one patient may be completely in denial, while another patient may be very well informed. One patient may be suffering from depression, while another patient may be completely in control and handling the situation with no anxiety.

So, it's important not to presume or prejudge the patient's mindset. Furthermore, both patient and physician have to understand that the management of the disease is a choice that requires teamwork and it's not something that the physician should be obligated or expected to do alone. A physician is responsible for his part, while the patient coordinates and communicates and plays the team player role. It is similar to a contract that the patient and physician make emotionally, spiritually, and legally. In other words, the more a physician tries to help a patient (beyond what is so-called expected in a normal fashion), and the more he puts his efforts and his knowledge into treating the patient, the more he expects the patient's trust, coordination, and compliance. It is unfair to expect that the physician put all his efforts into treating a patient who is noncompliant. It is important that from the beginning (or as soon as possible) the physician clarifies his expectations to the patient. If a patient is in denial, or is not completely compliant or on the same page as the physician, the physician must recognize this and address it as best he can and as early as he can, even, if necessary, referring the patient to a psychologist or an expert to coach the patient mentally and spiritually.

Not being on board and compliant with the physician's plan will significantly impact the patient's outcome, even with the best clinical care provided by the best physician. It is even the right of a physician to discharge a patient who is not compliant and it is the right of the physician to withdraw care if he knows that the lack of compliance will impact the patient's outcome. As such, he not only has to coach and instruct the patient, but also has the right to refer the patient to another provider if he cannot help them.

The excellence of care coming from a physician who puts the patient in front of his or her own family, the dedication and compassion that come from a physician who does everything he or she can for the patient, comes at a price. That price is the sense of gratitude and appreciation that comes from the patient. The physician who cares needs that positive reinforcement and encouragement to be able to continue that level of care. On the other hand, if the physician is bombarded with the patient's irrelevant questions, criticism, or doubts, the physician will certainly experience negative reinforcement and will go back to the standard of care to be offered to a patient with cancer, which we all know is minimal.

Going back to the contract that is built between the patient and the physician, although no paperwork or sense of gratitude is required as the price of extra effort and dedication from the physician, it really is the basis of a patient–physician relationship.

Going back to the core of the patient's goals and aims for their treatment, if the patient has a clear understanding of the goals and aims of the treatment and accepts those goals, there will be minimal dispute in this spiritual, emotional, and mental relationship. For example, a patient who comes to the physician with stage IV disease, based on the standard of care has less than 12 months to live. The patient certainly cannot expect a cure from the beginning. If a cure becomes a possibility, that must happen after the relationship has been built to the highest level, and after the treatment has proven to be effective for that goal. Otherwise, the goal of the treatment has always started with what expectations are defined historically. In other words, on the first consultation the physician must be careful not to provide false hope and must even be extremely conservative with the expectations provided to the patient. When I have a patient that I see from the beginning with expectations of survival defined historically between 6 months and 12 months, I tell the patient exactly that, and even emphasize that 6 months is what we are looking for, and we will try to maximize the patient's survival with the best efforts guided by the most research-based clinical practices' guidelines, and hopefully beat the 12 months.

But if the expectation from the beginning is beyond what the expectation is historically, the relationship cannot start. As time passes and the patient and the physician get to know each other better and their understanding of the philosophy of care becomes closer and closer, most of the time this positive reinforcement and encouragement continues to build a stronger relationship where the patient can pass the expected survival time, as the doctor puts more and more efforts into this relationship. The care provided by the physician to the patient is nothing but a relationship, and if the patient does not understand that, and starts the treatment and care based on an expectation that this is a duty and the responsibility from the physician; no relationship can be built on a sense of entitlement and such is true for the relationship between the physician and the patient.

Unfortunately, in the standard setting of oncology there is a significant barrier between the patient and the provider, starting from the goals of care to reporting side effects from toxic therapies.

It is important to re-emphasize that the provider's priorities at times do not match the patient's. The real question to ask is not "What is the matter with the patient?" Rather, it is "What matters most to the patient and her or his family?" The latter is where the most focus should reside and where energy should be spent.

For example, many surveys have shown that in ovarian cancer, patients with advanced disease care more about their quality of life than tumor response.[25] Oncologists have the worst reputation for reporting negative side effects of therapies (more than 50 percent are underreported). This was a survey of 1,090 patients with breast and non-small-cell lung CA, with underreporting of up to 74 percent of patients' reported side effects! This raises the question of the reliability of clinical trials assessment for providing a meaningful assessment at the individual patient level of the risk-benefit ratio associated with study regimens.[26] Also, oncologists tend *not* to discuss prognosis and survival with patients. It is reported that only 18 percent of oncologists discuss the prognostic estimates with patients. Not only that, but they only accurately identify 35 percent of patients with cancer who are likely to die within 6 months (!!), delaying referrals to palliative care and continuing futile therapies, for financial reasons.[27]

In countries where there is no defensive medicine practiced, and the physician-patient relationship is always based on trust, as the patient has minimum rights to file lawsuits for his or her care, the risk of negligence obviously is higher, yet the quality of care is higher too. This is because many times physicians go beyond what they are supposed to do, simply because they are not afraid of the patient or the possible meritless lawsuits. In the field of cancer this is even more important, especially when you are dealing with a cancer patient with a poor prognosis. The expectation historically is very minimal for a patient with stage IV disease who has a small chance of surviving. Because the disease is not curable, the patient automatically has lost his or her grounds for blaming the physician. Yet, the physician needs to step outside of the boundaries and receive the necessary courage from the patient to be able to provide care that beats the odds. Beyond just the sense of protection that he needs to receive from the patient, he also needs to receive the sense of gratitude that encourages him to do the best he can, to beat the odds for the patient who is not supposed to be cured. That said, personality traits are much harder to influence; it is more effective to teach coping skills.

In the end, we all should remind ourselves that bedside manners are, in the eyes of the patient, their very first impression of the care we provide to them. In a survey of 374 patients with breast cancer, patients' perception of the quality of their care was strongly associated with the "process of getting care" rather than the care itself or medical intervention.[28]

Another issue is the state of mind of the patient with advanced cancer. The majority of patients at this stage are depressed, and as such the 2014 guidelines from ASCO recommended routine screening for depression in cancer patients. There is also strong evidence suggesting that alternative medicine is effective in this setting.[29]

The element of depression, especially when patients are seen after receiving the standard of care, is extremely influential in the patient's ability to make decisions, to form proper judgement, and to interact with their providers as they are trying to provide a meaningful solution. As such, the expectations are higher for cancer providers; they must treat not only the patient's body, but also the soul, as both are harmed after toxic therapies.

Chapter Four

Goal of Care

In this chapter I want to discuss the philosophy of care. It is very important that every physician understands before attacking the patient's cancer that this is a fight that comes with a price. Regardless of the nature of the treatment, whether it is cytotoxic treatment, conventional oncology care, radiation surgery, chemotherapy, or even purely natural treatment, there are side effects. Some of these side effects may be positive and some may be negative. Thankfully, one of the major successes in providing natural treatment is to improve quality of life to some degree in certain patients. Those are positive side effects—for example, the patient's energy level, stamina, and mood can improve with certain natural treatments. Yet there is a balance between quality of life and the antineoplastic effects of certain therapies. For example, a natural treatment can improve the patient's quality of life but at the same time can have no impact on the tumor, or can even make the tumor grow faster. Other natural treatments provide better quality of life without negatively impacting tumor size; in other words, they are simply guided to improve the patient's energy, stamina, and immune system without having any effect on the tumor.

The best natural therapies are those that not only improve the patient's quality of life, but also have direct antineoplastic effect. You can appreciate that, depending on the aim of the treatment and what the provider is trying to accomplish, he can choose between different natural treatments, either alone or in combination with certain categories of conventional treatments.

So, it is important that the expectations of the care are in line with the philosophy and the style of practice that are designed from the beginning.

For example, some alternative practices or clinics only improve the patient's quality of life. They try to minimize the side effects of conventional treatments and they do not claim or try to accomplish any antitumor effect. Other clinics, where the philosophy of care is to provide the least toxic and most effective therapies, sometimes view the toxicity as acceptable because an antineoplastic effect is expected at the same time.

Now, when a patient comes to see me, depending on what protocol is used the philosophy of care is designed around minimizing any negative side effects and maximizing all of the positive side effects and the antineoplastic effect.

This understanding comes with practice. The more a physician practices different protocols the more he understands the variety of patients' responses and reports and engages with the treatment accordingly to accomplish minimum toxicity and improve quality of life. Despite the nature of most antineoplastic therapies, there is no set protocol that never adapts to the patient's feedback; in other words, even with established treatments that could work on reducing tumor size and maximizing cancer response there has to be room for adaptation depending on the patient's feedback to every protocol. This adaptability is extremely important (and most appreciated) in the settings of integrated care, even though it is barely present in the conventional world. Some patients may have minimum pain tolerance and as such even a venipuncture could be a problem.

Other patients may be more sensitive to certain products or even be allergic to them, so the most conservative way of treating cancer starts with assessing the patient's medical history and their perception of side effects. If the patient's perception of pain and discomfort, or if their perception of investment, either financially or timewise, is not in line with the physician's, the patient must be either discharged and referred to another provider or has to be coached to be on the same page with the physician. Some patients go to alternative practitioners simply because they do not want to experience any side effects; that is why they refuse chemotherapy and other conventional treatments. As such, these patients are not good candidates for alternative therapies. Some of these therapies could come with the price of some negative side effects, which, although they are not significant and are reasonably manageable, could be problematic if a patient's perception of alternative therapies is simply therapies that improve quality of life.

In my practice, I have learned how to manage the philosophy of care based on the least toxic and most effective treatments with the majority of my patients. I explain the expectations for the treatment from the outset, so we are all on the same page, but the more trust is built, and the more accountability is appreciated from both sides, the better the relationship grows and the more effective the management becomes. Finally, I want to note that quantity of life and quality of life are two completely different things. For a patient to live longer they must not sacrifice their quality of life; in other words, the price of living longer should not be to live miserably. This is a set standard goal for the practice of integrative oncology.

Finally, I would like to discuss the financial aspects of integrative oncology practice, one of the main criticisms about any nonstandard therapy. It is a fact that most insurance companies deny payment for these so-called experimental treatments. Physicians who practice outside the box and provide therapies that have been scientifically studied and have strong scientific rationale and merits should not be afraid that the financial burden on the patient will reflect on the quality of their care. Also in my view, patients should not carry the burden of being unable to pay for their cost of care. It is completely illogical, unethical, and inhuman.

None of the practitioners who follow my recommendations should make finances the priority of their care, and should always remember that interestingly, the patients who accomplish the best results in my practice are the patients who are assisted financially.

Finances are secondary. Number one, if the physician puts all their effort into providing a meaningful outcome and positive results, generally speaking, patients find a way to pay. When they see results, especially in a short period of time, they're encouraged to continue the plan of care and find resources wherever they can to pay for the cost of care. It is not unreasonable to provide discounts to certain patients, but it is more meaningful to reach out to donors who might want to sponsor a group of patients who cannot afford to pay. Some of the most generous donors are former patients who were among the best responders to the treatment itself. I'll give you an example. Ten years ago I treated a famous, wealthy baseball coach who had mantle cell lymphoma.

Mantle cell lymphomas are characterized by reciprocal translocation (11, 14) and possess the unfavorable characteristics of both indolent and aggressive B-cell lymphomas.

In general, treatments of B-cell lymphomas, mantle cell lymphomas, and chronic lymphocytic leukemia now overlap at times; for example, these patients can mostly benefit from bendamustine and rituximab if chemotherapy is chosen, or Bruton tyrosine kinase inhibitors, and of course undoubtedly from epigenetic therapies. All three groups of patients (i.e., those with Waldenstrom's macroglobulinemia, CLL, or mantle cell lymphomas) benefit from similar therapies. (For example, the standard of care for CLL is BTK inhibitor in both initial and relapsed patients, in combination with rituximab in initial and in combination with bendamustine in relapsed).

In the case of the baseball coach, he had already failed chemotherapy. This patient responded in a matter of four days to the epigenetic protocol I offered him. As a result, the patient sponsored a group of others who did not have the financial means to afford the treatment. Twelve years after his diagnosis he is alive and cancer free. (No cure for mantle cell lymphoma has *ever* been reported, especially with high Ki-67, in this case 90 percent.)

If I had thought about just finances from the beginning and not about the results, this would have never happened, but because I put all my efforts into making every case the most successful, I beat all the odds, and the patients appreciate the efforts and find ways to pay not only for themselves but also for others.

I've had patients whose care I've sponsored all myself and they paid me nothing, but in reality, they paid me a big prize, and that was their sense of gratitude and the fact that I could use their information for the next patient that I would treat in my practice; this was priceless. So, at the end of the day the results speak for themselves; if a provider can prove that his results are better than the standard of care, there is no way that this provider will struggle financially, and the finances are surely guaranteed.

Here, I want to present some difficult cases and discuss the challenges in relation to their care.

Case #1 A 67-year-old female with recurrent stage IV ovarian cancer with peroneal carcinomatosis who had failed platinum-based chemotherapy regiments, meaning that after the first treatment, the tumor had recurred in less than 6 months. The patient was referred by another physician to my practice and initially was under the impression that she could find a cure. Financially the patient was in a good position, but because of her significant hesitations about so-called alternative therapies (based on the cash-based nature of the therapy) she was reluctant to pay for her care. Yet, the patient expected a cure and gave examples of friends she had known in the past who had lived with ovarian cancer for 10 years or more.

How would you approach this patient in this scenario? Number one, you could discharge the patient either from the beginning or after counseling her, based on the conversations showing that the expectations are not on the same page, and the patient is either not able or not willing to pay for the cost of care.

Number two, you could treat this patient to the best of your effort, regardless of the difference in expectations and regardless of the financial means.

Number three, you could start the patient on treatment after extensive counseling and further evaluate the patient for her expectations and finances.

Number four, you could encourage the patient to see another provider and provide a referral as well as evidence in regard to your inability to cure this patient at this stage.

You can see that these answers are somewhat similar, and basically open further options for the provider to explore. None of these options is wrong, but their sequence is important. For example, if a provider discharges this patient without trying to treat her and without investing time in her, that is not wrong. If the patient is discharged from the beginning she will have time to think about the expectations and she will have the chance to reconsider her expectations and come back. Also, it is not wrong to start the patient on treatment after coaching her and waiting to see the results before you dive deeply into the prognosis. My preferred approach with this patient is to keep all options open, meaning that I would give her a chance to change her attitude in regard to her expectations. I would explain the fact that I cannot cure her from the beginning, yet I would not close the door. I would also try to assist her financially, as much as I can, but also give her time to understand that she has to come up with financial means to be able to proceed with this approach. My approach would be to also give the treatment a chance to start working. If the patient responds very well to the treatment that I propose, I would encourage her to continue and give her more hope. At the same time, if the treatment does not show superior results, I would emphasize the fact that I may not be the best person for the patient to see. I would keep the discharging option for myself open, but I would do that at the time that when I saw minimal response and maximum expectations from the patient. But if her response was growing and her expectations are coming down, my hope is that they will reach each other at one point.

So, after a few weeks I will know if the patient's expectation has reduced and if my expectation for the patient to have better results has improved with the treatment. I would work with her through this process financially so she could at least afford an initial course of treatment. As you can appreciate, this is a difficult process, because I am dealing with a difficult patient, a patient who does not recognize the severity of her disease and has very high expectations. In addition, the patient is in financial trouble and yet has a very advanced disease, but still I will always give a chance to the patient, to myself, and to the treatment.

Chapter Five

What Is Cancer? (Enrichment of DNMT3A Mutations?)

Addressing Needs for Young Adults with Cancer

In the US, in 2021, almost 1.9 million people were diagnosed with cancer (new cases), and 610,000 died of the disease. Currently we have 17 million people with cancer in the US alone (projected to increase to 22 million people by 2030). In China there are 4.5 million new cases of cancer and 3 million deaths reported annually. Globally, over 10 million people die of cancer every year. *That is one death in less than every three seconds.* In medieval Britain the cancer death rate is estimated to have been only 14 percent, when now it is 40–50 percent.

In 2021, the 5-year survival rate was highest for prostate cancer, melanoma, and breast cancer, and lowest for pancreatic cancer, liver cancer, esophagus cancer, and lung cancer (regardless of stage), but this is only reflecting the mass incidence of diagnosis of these cancers in the early stages, versus other types that usually are diagnosed later in the game.

Incidence of cancer may have decreased in certain populations, but the rate of death has increased in others. For example, between 2001 and 2009 the survival rate of patients with metastatic melanoma was higher compared to 2009–2014. The incidence, though, has declined for both lung cancer and melanoma, whereas the rates have increased for prostate, colon, and breast cancer, especially in younger individuals. Cancer incidence in women in general has increased (an annual average percentage of 0.2 percent). The incidence also increased between 2014 and 2018 in all children, adolescents, and young adults of all races except Native Americans.

Globally, lung cancer has a fatality rate of 32 percent, prostate cancer 30 percent, central nervous system cancer 27 percent, and breast cancer 10 percent. Colorectal cancer has passed leukemia as the second leading cause of death in men aged 20–39. If we add colon, breast, and lung cancer cases in women, and if we add prostate, colon, and lung cancers in men, we have covered 50 percent of all cancer types altogether.

The sad news is that the death rate for certain types of cancer has actually increased globally. This is specifically true in the case of breast cancer in women aged under than 40 (it increased in 2020 by a rate of 2.8 percent). Also, in colorectal cancer the early onset type of the disease has significantly increased since a decade ago, especially in the Middle East. (More than half of the cancer cases are reported in patients younger than 50 years old.)

In the US this number used to be 12 percent of patients. For this group, the incidence and death rate increased by 1.3 percent and 2.6 percent respectively between 2012 and 2016, but in 2021, the rate increased to 50 percent for patients with colorectal cancer under 66 years of age. Also in 2021, for the number of women presenting with stages III and IV breast cancer, the incidence increased from 12 percent to 18 percent. As we now know, the prevalence of HPV/EBV+ noncervical cancers is increasing in the United States (anal, nasopharyngeal, gastric, etc.) despite all of our advanced therapies. We also know that the rate for thyroid cancer in children increased (doubled) between 1973 and 2013.[30] It is possible to think this could be caused by radiation to the neck (as part of treating Hodgkin's lymphoma or others), iodine deficiency, or smoking…when treated with radioactive iodine, you can also induce secondary malignancies.

The reason behind the increased rate of cancer incidence in the younger population is simply accelerated aging in the new generation, as someone with a calendar age of 35 can have a biological age of 50. Different cell DNA (cDNA) methylation patterns as well as telomere size can show the biological age of an individual. Chemotherapies activate the MAPK and induce mitochondrial dysfunction, causing aging. (As an example, a daughter of my patient with AML wanted to be screened for cancer-causing germline exposure. The test results indicated the presence of two single nucleotide polymorphisms, both involved with early aging, TERT, and NBN. This shows that this person is susceptible for early leukemia, as her father was, at young age.)

In general, the risk of positivity for a germline mutation is almost 100 percent when you have a parent or sibling with cancer, as studies have shown that for the most common moderate penetrance cancer genes (defined by lifetime cancer risk of 20 percent or more), such as ATM, CHEK2, or PALB2, the risk is generally similar to that experienced by a person whose sibling has been diagnosed with cancer (breast CA in this case). A perfect example in my clinic is a patient with breast cancer, who had three siblings who had died of cancer (pancreatic, breast, and ovarian). The tumor genomic confirmed PALB2 as the genetic driver and the germline testing confirmed the presence of this alteration.

Unfortunately younger individuals with cancer do not respond to the traditional therapies the same way. For example, it is reported that adolescents and young adults with cancer who are diagnosed between the ages of 15 and 39 comprise 5 percent of cancer patients in the general population and yet there is *no* reported improvement in their survival in any type of cancer. The disease in this group of patients simply has a different biology; specifically, breast, bone, cervical and ovarian cancers have increased.

It is reported that hematological cancers such as ALL reported in this group of patients respond better to protocols developed for children, not adults. There is a significant barrier for this group to access clinical trials because of the arbitrary age cutoffs. In general, patients in this group have an unfavorable genetic markup (such as triple negative features in breast CA, or BCR-ABL features in ALL) significantly impacting their survival, specifically in colorectal and uterine cancers. As such, germline testing is mandated for this entire group of patients. Recommendations such as double mastectomy are made more often due to germline mutations such as BRCA or TP53 (which makes radiation possibly harmful).

There is a high presence of germline mutations (about 17 percent) in this group of patients (less than 40 years old), but there is also a ninefold increased risk of local recurrence in this population. This is important, as it speaks loudly about the fact that risk of recurrence is not explained by germline mutations at all, because if this was the case, there would have been an increase of less than 20 percent and not a ninefold increase. Pretty much what the scientists have missed is the fact that the cause of the initial cancer is not germline mutations, and the same concept applies to recurrence of the disease. Many studies have now confirmed that constitutional epimutations drive carcinogenesis in young adults.

Another perfect example is the frequency of germline mutations in both smokers and non smokers with lung cancer. There is absolutely no difference in germline mutations in smokers and nonsmokers, which again emphasizes the epigenetic role in the disease incidence.

The application of chemotherapy also needs to be questioned in this group, considering the fertility preservation choices and patient desire for that. In general, premature ovarian failure happens in about 13 percent of patients receiving anthracyclines. Modifying the standard of care protocols are justified in these scenarios., For example, a single agent, Taxol, can be used in combination with HER2 inhibitors for patients with HER2+ cancer, if the tumor is less than 2 cm (and no regional lymph node identified). Avoiding oopheretomy is something that needs to be considered cautiously, and using luteinizing hormone agonists instead could preserve the ovarian function.

It appears that the positive effects of chemotherapy are contributed only through ovarian suppression, and as such using chemotherapy can be altogether avoided if ovarian suppression is planned. Overall, neoadjuvant chemotherapy is only recommended if there are high-grade tumors, with an Oncotype DX score of over 25; otherwise this approach may not benefit the patient and certainly takes away the opportunity to receive the full genomic information after surgery. Also, using tamoxifen can cause premature ovarian failure in addition to its teratogenic nature (i.e., it requires contraception if used).

All five classes of gene expression labs are useful in predicting chemotherapy benefits: Oncotype score (16 genes), MammaPrint (70 genes), Breast Cancer Index (11 genes), EndoPredict (12 genes), and Prosigna (50 genes), although in my practice I have used only the Ki-67 marker in pathology recently, as I have seen the most meaningful correlation with the Oncotype scoring, which used to be ordered for every newly diagnosed early stage patient with breast cancer in my practice years ago.

Breast cancer during pregnancy and postpartum (5–10 years after pregnancy) represents about 40–50 percent of breast cancer cases in women younger than 45 years of age. It is speculated that the regenerating breast tissue after pregnancy, a phenomenon called "involution," contributes to scar formation and the potential of metastasis. That said, termination of the pregnancy has not been shown to improve survival.

One of the main genomic characterizations of early onset cancer is the presence of NOTCH1 and TGFBR2 (this is especially noticed in the Black population). The epigenetic influence of carcinogenesis is driving the tumor incidence in younger individuals with or without germline mutations. The TGF-beta, SMAD, and Wnt mutations are increasingly seen with PI3K CA activating mutations, and in fact the TGF can independently affect the PI3K. TP53 and APC do not have this pattern. The main concern about using any TGF blocker (Galunisertib by Eli Lilly, TEW-7197 by MedPacto, and NIS 793 by Novartis, which is an antibody suggested for pancreatic cancer to inhibit fibrosis) is the cardiotoxicity, as TGF physiologically is associated with cardiac remodeling.

In this chapter I would like to summarize the most clinically relevant information in regard to the definition of cancer, and how the definition has changed in recent decades, interfering with the way that we do research and in the way we treat patients. Obviously with the new understanding of cancer stem cells we realized that behind-the-scenes players are the true indicators of a patient's response to any treatment over the long term. As such, cancer stem cells are identified as key players in the induction of cancer (or what we call carcinogenesis) and the role that these stem cells in the local area play in a true repair system.

As a whole, the previous common thought was that cancer was a byproduct of genomics and environmental risk factors. In my opinion, it is simply irrational to split the two. In fact, there is no line separating genomics and epigenomics, except time. A good example is a study that showed that lack of sleep causes DNA damage through dysfunction of the DNA repair genes, and causes cancer.

Some have also argued that inflammation causes cancer, and yet we know that the presence of ulcerative colitis by increased IL-17 and TLR signaling could well be promoting cancer development, only when an epigenetic influence dictates as much. In AML, we also know that inflammation has been blamed for causing leukemia (through IL-1) if the PU.1 is impaired. PU.1 is a break in carcinogenesis when there is inflammation (it causes quiescence and limits hematopoietic stem cell expansion).

Therefore the cancer mutations are *plastic* in their penetrance and only manifest themselves when the stroma is in favor, and there is a green light on the epigenetic level. This concept is validated by whole genome sequencing showing that DNA alterations extend from the homologous recombination genes at a point mutation.

This theory was also upgraded to include the epigenetic role, before we realized that genome is the same as epigenome, only with a matter of time difference (all genetic abnormalities can be caused by epigenetic aberrancies). We used to say that methylation causes mutation; I believe methylation is indeed mutation, only with a time difference.

In other words, when we talk about BRCA genetic mutation, we are only looking at a "product" of epigenetic variance of this gene by a methylation process that started long before the gene was mutated. This concept of conversion from energy (epigene) to material (gene) and vice versa is the particular area of my interest. Based on Einstein's theory of material (M) and energy (E), both M and E are convertible to each other, and one can be the cause and the other the effect. So if E is the cause for M, the conversion of M to E would be a matter of time, and vice versa. In the world of epigenetics, the conversion of E to M is best explained in the hereditary fashion of the diseases caused by epigenetic abnormalities; for example, for someone affected by lung cancer with genetic mutation at KRAS, the smoking is the main cause. That said, this person may have never smoked, yet still be affected by lung cancer. When questioned about their family history, 100 percent of the time the patient can recollect a family member, either a parent or grandparent, who was a heavy smoker. This phenomenon is widely known in epigenetics, when next generations are affected by an accumulation of events that have only been passed through epigenetics. This passage can be through 10 generations in an individual.

Now, if we take out the time element in between, the cause and effect become the same. The example that I always give is when I let my pen fall to the floor by gravity. The cause is the pen left in the air and the effect is falling; the only difference is time. Here I am trying to emphasize the relationship between epigene and gene, when one is the cause and the other is the effect.

Going back to the BRCA story, this also explains why the incidence of breast CA in an individual with BRCA positivity is 72 percent (in females, and 10 percent in males) in this century, compared to only 26 percent in the 1940s. The same concept applies for serous ovarian cancer with a BRCA mutation (mucinous types have TP53 mutations, not BRCA).

For men with prostate cancer, 5 percent carry a BRCA2 mutation, and about 1 percent a BRCA1 mutation. Men with breast cancer have about 10 percent BRCA mutations, and BRCA+ individuals have 10 percent incidence of breast CA. This is also true for women, as the total germline mutations is about 10.4 percent in women with breast cancer (in comparison to 13.3 percent reported in the general population for all types of cancer. BRCA/CHEK2/PALB2/ATM in order of frequency). Twenty defined breast cancer-associated genes have been identified: BRCA, ATM, CHEK2, PALB2, TP53, STK11, RAD51D, MLH1, MSH2, MSH6, CDH1, BRIP1, BARD1, EPCAM, NBN, SMAD4, NF1, PMS2, PTEN, RAD51C.

Pancreatic cancer patients have about 5.5 percent BRCA positivity. BRCA positivity also carries the risk of leukemia, especially in patients after receiving chemotherapy. In pregnant mothers with positive mutated BRCA, genetic testing is recommended at gestational ages 12 to 16 weeks to rule out the presence of Fanconi anemia in the fetus (Fanconi syndrome is biallelic mutation of BRCA1, or both parents pass the BRCA2/1 mutated genes to an offspring). Annual screening with MRCP and endoscopic ultrasound is recommended for these patients. I wonder how someone with BRCA mutation can be convinced to do 20–30 annual MRCPs after the age of 50?

Genome wide association studies have significantly increased our ability to connect the presence of SNPs with prediction of biology. For example, in multiple myeloma, MGUS and SMM, 3p22.1, 6p21.33, and P154.3 have been associated with the incidence of disease. At the same time, we know that people who have been exposed to Agent Orange and pesticides, as well as 9/11 rescue operators, have significantly higher risk of the disease. This explains why the incidence of MGUS is nearly 15 times more in the US than in Asia (6 percent compared to 0.4 percent). It is certainly predicted that the exposure to these environmental factors has been the initiator of transcriptional activity of already existing aberrant genomics, or what we now call epigenetic coding (identified in Epigenomic Association studies). Another example is the SNPs related to risk of estrogen receptor breast CA in BRCA+ individuals (ZNF423, CTSO).

The only epigenetic drug approved for use in MM is panobinostat (used alone or in combination with mTOR inhibitors). This drug is an inhibitor of HDAC3 and 6. HDAC3 is involved in regulation of cellular signals, apoptosis, PI3KCA and MEK, downregulation of pro-survival factors (MCL-1), and upregulation of cell receptors/ligands (CD38). HDAC6 is in charge of inhibiting the heat shock protein 90 chaperone and tubulin. HDAC1 is involved with TP53, RUNX3, and AR. The HDAC family also controls c-Myc, HIF, MyoD, SMAD, STAT, and sirtulins.

The process of transition from a precancerous to cancerous lesion or disease is another example of this carcinogenicity stepwise process. Again, as an example, MGUS nearly always precedes MM, and the same is true for DCIS and breast CA. Thus:

$$\text{MGUS} \rightarrow \text{MGC} \rightarrow \text{SMM} \rightarrow \text{MM.}$$

(The paraprotein in serum, for example, can increase from 1.5 gr/dl to 3, and clonal plasma blasts increase from less than 5 to 10, and the diagnosis changes, but again this is a "process" before a patient is actually diagnosed with obvious CRAB syndrome of MM).

The cytogenic abnormalities identified will *only* be able to cause the disease in the presence of favorable epigenetic influence. As such, the aging of stem cells and "accumulative" effect of collective aberrancies over time explains the reasoning for increased risk of progressive disease, from MGUS to MM, by a ratio of 10 percent per year. This aging process is simply the epigenetic aberrancy explained here. The standard of care in this scenario provides *no* meaningful solution because it only recommends monitoring a patient with this scenario with a bone survey every three months to look, after the fact, for bone-destructive lesions.

This is a good opportunity to discuss the aging process, as it applies to the cancer therapies themselves. Chemotherapy and radiation are notorious for causing accelerated aging in patients and in their cellular processes.

For example, it is estimated that both children and young adults receiving traditional therapies have an increased biological aging of 10–20 years depending on the agents used (anthracyclines are twice as bad as taxanes). The telomer size and DNA methylation assays have confirmed the magnitude of accelerated aging in these patients, many of whom, especially children, will also suffer from clinical deterioration of their intelligence for a long time after chemotherapy, called chemo brain. Survivors have a markedly increased risk of developing comorbidities, frailty, *and cognitive impairment*. The results of this negative impact are *irreversible*. In these patients, by the time they reach age 30, nearly 35 percent have the clinical phenotype of a person 65 years old, as evidenced by dramatic increases in their incidences of cardiac disease and new secondary malignancies (on average at 23–27 years). This is also true for bone marrow transplant with 34 years accelerated aging. Radiation causes sixfold increased p16 in biopsies taken from the scalps of children exposed to cranial radiation. In breast cancer, post chemotherapy patients experience about 15 years advanced aging, measured by p16.

As such, it has been shown in several studies that chemotherapies can cause microbiome changes in children that are linked to heath implications later in life, primarily obesity, asthma, diabetes, and even schizophrenia. These effects were caused earlier, through mitochondrial disruption caused by the cytotoxic effects of the chemotherapy. Sarcopenia, cachexia, and frailty all are caused by either cancer, or cancer therapies, through the process of inflammaging. It is also important to know that the accelerated aging caused by chemotherapies happens through changing the epigenome landscape. The methylation patterns of DNA allow us to distinguish the biological aging very accurately, using a GrimAge, extrinsic epigenetic age acceleration, or phenotypic epigenetic age acceleration test. Also, inflammatory markers such as IL-6, TNF-alpha, IL-1B, CRP, and D-Dimer, can correlate with this process. The good news is that the epigenetic clock is amenable to deceleration with different strategies, like vitamin D, supplements, and a thymus rejuvenation protocol (using omnitrope, DHEA, vitamin D, and metformin).

Research has shown the role of p16 expression as a robust marker for biological aging, in both T cells as well as normal tissues that mirror the T cell expression of p16. The higher p16, the less cellular reserve for oxidative stress and tolerance to chemotherapies, causing frailty syndrome with poor quality of life, common illnesses, similar to young adults with cancer. P16 is found on chromosome 9p2, and is a CDK 4/6 inhibitor that promotes cell senescence by preventing phosphorylation of RB1, leading to G1 cycle arrest. Its expression doubles in T cells, by every 8 years. Other biomarkers include beta-galactosidase, p16INK4a, TP53, PAI-1, p17, upregulation of cytokines, and growth factors.

This finding argues against the common thought that cancer is a killer. In fact, it suggests that the treatments of cancer are killers, as the more we treat patients successfully using toxic therapies, the more we impact the survivors' quality of life negatively!

The relationship between the stem cells and daughter cells after a patient is diagnosed with cancer is very unique. The stem cells, or mother cells, are differentiated by their nature; they're plastic, and they can commit to different lineages of cells, as such explaining the tumor's origin. They do not respond to the traditional ways of treating cancers, such as chemotherapy, radiation, and surgery. This leads us to question the reasoning behind radiating a prostate original tumor in a metastatic setting, or when after definitive therapy, the PSA rises above 0.2, not considering the negative side effects as described.

This is because they have mesenchymal origins, meaning they have been induced by local stem cells to attract bone marrow stem cells. They do not originate from the disease area itself; tumor recurrence has been reported even after 25 years of remission. The plasticity of these stem cells is controlled under the process of genetics and therefore although it is hereditary it is a process that is reversible, meaning that stem cells can be less or more stem depending on the epigenetic appearances, and because of this dynamic process all drugs so far have failed to target cancer stem cells effectively without damaging normal stem cells.

What clinically a physician has to understand and explain very carefully to the patient is the difference between the tumor and the cancer. The tumor is presenting daughter cells and the cancer is presenting stem cells. The tumor can be removed, so the patient can accomplish their mission, but the cancer stem cells are never removable. For a tumor to be detectable, in the case of breast cancer for example we need to have a minimum of 500 stem cells in the area to produce the colony of millions of daughter cells between the stem cells and daughter cells. Also, there are cells that are not totally stemmed but they're not totally differentiated from daughter cells. Both of these are called pleiotropic cells. Pleiotropic cells are cells that can be identified in a blood sample, in what we call cancer circulating tumor cells. In the most difficult to treat types of cancer, traditional treatments (surgery, chemotherapy, and radiation) have a significant role in the induction and plasticity of stem cells.

The activation of stem cells is mainly explained in the bivalency theory of Stephen Baylin, manifested by overexpression of c-Myc. in CTC; this explains not only the presence of a primary tumor in one person, but also the presence of a secondary primary tumor(s) in the same person. For example, in different stages of life, a person is diagnosed with several types of solid tumors, or a patient with MGUS can progress to have CLL, non-Hodgkin's lymphoma, and MM by a very similar pattern and similar epigenetic switches. This is very relevant for epigenetically driven tumors, such as lobular breast CA or hematological cancers. This phenomenon is also seen in CLL patients when diagnosed with skin cancer, which usually has a poor prognosis.

For the CLL to happen, the chromosomal abnormalities are not enough; we need other mechanisms involved. One is called MLL, or the abnormal (loss of function) of methyl transferase, a genetically dictated enzyme, which is seen mostly in Asians and is called KT2D/B or MLL. It is seen in ALL, AML, and even CLL, reported by my clinic for the first time in a non-Asian woman. The presence of epigenetic abnormality is associated with a high risk of chromosomal abnormalities (such as deletion 11 or 17) and TP53 mutations. Although a poor prognostic marker, it is very effectively modifiable by histone deacetylase inhibitors. These cases usually have high transition to Richter's transformation lymphomas. Richter's transformation commonly is a result (65 percent) of chemotherapy use, whenTP53 and NOTCH1 are activated. The current standard of care addresses transformed lymphomas the same as diffuse large B-cell lymphomas; as such, chemotherapy (R-CHOP) is recommended—amazingly, knowing that this was the cause of the transformation to begin with! My recommendation on BK inhibitors stays the same in this scenario. The mechanism of transition from indolent lymphoma to aggressive lymphoma always involves the PI3K CA pathway, and as such using HDAC inhibitors *must* be considered.

There are studies on a gene-editing technology, CRISPR technology, that would remove the IGFBP3 protein that has been blamed for its role in MLL leukemias. IGF2BP3 has been overexpressed in about 15 percent of different solid tumors.

There is only a 7 percent response rate in large B-cell lymphomas to salvage chemotherapy, and patients with refractory disease only live on average for about 6 months.

What is extra chromosomal DNA? Chromothripsis is a process of chromosomal shattering caused by chemotherapy, which causes DNA glue by attaching to the DNA strand breaks that occur during treatment with chemo and making new double-stranded DNA. As such, it is always recommended to add DNA repair inhibitors to the cytotoxic medications to prevent this process.

We also know that TP53 mutated clones emerge after using chemotherapy, triggered by DNA damage causing tumor clonal expansion with higher resistance. These mutations can be associated with CDKN2A and RB1. This is an important association, as in my clinical experience, it potentially creates a treatment option that includes sequential inhibitors of mTOR with cyclin-dependent kinases, along with the epigenetic modifiers (DNA demethylators) used to unleash the RB1. The combination is vital, as otherwise the cyclin-dependent kinase inhibitors will induce resistance by mutated RB1. This phenomenon is commonly seen in breast cancer, when a patient develops resistance to this class of drugs. This resistance can be inhibited and the durable response observed when a combination of epigenetic drugs are used. The association between both Li-Fraumeni syndrome and hereditary retinoblastoma and osteosarcoma makes one think the mutations of TP53 and RB-1 are somehow linked.

Induced pluropotential stem cells are the mainstay of certain therapies. It is important to know that the presence of c-Kit is a hallmark of mesenchymal features for stem cells, before it is differentiated to adipose cells. There are certain microRNAs (MicRNA 181) that are involved with this process of progenitor state of cells.

The three genetic markups of OCT4, SOX2, and Nanog are correlative with TGF, FGF, and Wnt activation in stem cells. FGF activates Akt, TGF activates STAT, and MAPK and Wnt activate GSK (lithium inhibits that). Further epigenetic modifications also control the activation of three hallmark genes (OCT4, Nanog, and SOX2). TGF-beta also drives the metabolic dysfunction of natural killer cells. As such, scientists at MDA have been able to remove the TGF effect on NK, by using CRISPR technology to modify its receptor, causing augmented NK activity.[31]

The only approved drug for TGF is luspatercept for treatment of MDS. The concept is that clonal hematopoiesis happens as a result of epigenetic disruption, which further causes cytopenia of unknown significance, which further can develop into MDS and AML (30 percent of the time). The identification of at least one mutation at mutated allele frequency of more than 2 percent is diagnostic. If this test (which we order through a genomic lab) is not available, the cytopenia that has been ruled out for other causes, if it lasts for over 4 months, is suggestive of this disorder (CH). My rationale for a therapeutic approach at this time would be to measure serum TGF-beta 2, and if increased, use epigenetic therapies to lower it.

The mechanism of action for luspatercept is to inhibit the SMAD 2/3, to reduce the need for transfusion for at least 8–12 weeks. Another TGF inhibitor is bintrafusp alfa, a bifunctional fusion protein that targets TGF-beta and PD-L1. It is speculated that resistance to immune checkpoint inhibitors is related to growth factors in the microenvironment, including microbiomes—clostridia mainly—and as such proton pump inhibitors are not recommended to be used along with these drugs.[32]

There are other bifunctional proteins in discovery; for example, one that binds the prolactin receptor on the cancer cell's surface and activates the natural killer cells has been studied by Yanzhang (Charlie) Wei, biological sciences professor/associate chair at Clemson University.

There was a recent study that actually showed using metronomic Cytoxan (at doses equivalent to 100 mg a day in humans) in combination with sodium bicarbonate orally caused complete cure in animals with triple negative breast cancer, treated with tumoral ablation. This concept significantly changes the way that we can look at TNBCA, as it does not include any standard chemotherapy, and it really targets the microenvironment.

Colon cancer: TGF→APC mutation→DNA hypomethylation→KRAS mutation→TP53 mutation→ cancer**→ EMT activation by Wnt/Snail and Slug through histone demethylation→ Metastasis

**= Wnt activation (TP53 activates Wnt), in the presence of favorable epigenetics

Snail is recognized in stroma and luminal cells in breast CA and other types of cancer, and Slug in basal cells. Slug generally is a sign of activation of the stem cells and Snail is a marker for the stemness of these cells. ZEB1 also plays a role. (This is Epidermomesenchymal transition.)

EMT is a way of reproducing the new progenitor cells. TGF, Snail, and SOX are all markers for these mesenchymal cells. There are 20 drugs screened among 16,000 compounds that can reverse the process. For example, salinomycin and abamectin are two antihelminthic compounds that have these properties. Forskolin, on the other hand, is a natural plant extract with the same property that interestingly also inhibits the protein kinases in CLL. Ibrutinib effectively inhibits c-Myc, as we mentioned earlier, as well as glutamine. Forskolin also inhibits ERK, causes apoptosis, Bcl-2, and PARP, and has been investigated in multiple myeloma along with dexamethasone. It increases the cAMP. It has been studied in TNBC and AML, as well.[33]

One biomarker for EMT is microRNA 205, which lowers EMT. AGTR-1 gene editing also has been shown to be effective in EMT inhibition.[34] Losartan technically is supposed to do the same.

Although there are a few stem cell drugs being investigated (for example, hedgehog pathway inhibitors such as glasdegib have been suggested for basal cell carcinomas and AML), the toxicity of these drugs is overwhelming. Cemiplimib is the immune therapy approved for BCC after hedgehog inhibitors fail. The best I have used for this disease is Vaniqa (this cream is approved for skin hair removal, but the compound eflornitine is approved for in Africa for treatment of sleeping disease). This compound was discovered in Germany for treatment of cancer in 1920 (before the inventor gave up his patents to the CDC!!). There also are studies supporting the use of HDACi in topical form for treatment of basal cell carcinoma, refraining from surgery and other toxic therapies.

The copy number of genetic variations (copy variants such as genetic doubling) is a good early signature, but even that is now being questioned, as we know the higher the copy number of the genes, the higher the ability to tolerate mutations without cell death; therefore, cells with lower genetic doubling will not survive.

Questioning the origin of the tumor: It is predicted that "clonogenic precursors" are origin or pediatric tumors. In other words, the theory of accumulative mutations that happen in DNA is now under question, as normal cells have demonstrated the same mutations in the same organ but outside the tumor itself. This clonal expansion is caused by hypermethylation of H19 in the case of RCC. Since the stem cells are the origin of the majority of cancers, it is specifically important that epigenetic therapies are applied sooner than later to induce differentiation. This is particularly important in pediatric malignancies, where the fetal cells are the origin of the tumor. What are called preleukemic clones are the initial clones that the newborn carries upon birth, which mostly never develop leukemia, unless a second mutation is triggered by a biological event, generally as a result of an inappropriate immune system reaction to an infection. There seems to be a theory on the increased sensitivity of children towards infections when there is minimized exposure to common infectious agents; in certain families that avoid such exposure, when the first exposure happens later in life it may cause an abnormal reaction and causes a second insult, causing ALL. There also is an assumption that a planned C-section increases the risk of ALL.

Hereditary cancer syndrome: It is known to physicians that about 1 out of 8 cancer patients have a genetically inherited form of cancer; that said, in my view all cancers are driven by an epigenetic abnormality and related genetic mutation. Basically, only 1 of 8 are mutated forms we can test for. That said, all the genetic tests commonly practiced in mainstream clinics, such as BRCA, ATM, and so on, miss 62 percent of genetically inherited cancers.[35]

So for example, 10 percent of gastric cancers are genetic, meaning that there is a germline mutation identified at CDH-1 or CTNNA1, but that said, patients who have a family history of gastric cancer *or* lobular cancer are at increased risk for gastric cancer. Why lobular cancer? Well, simply because lobular cancer is a type of breast cancer that is 100 percent epigenetically driven, meaning that the same concept applies to the gastric cancer. In fact, you can see both gastric cancer and lobular breast cancer in the same person at different stages of life. This concept explains the increased incidence of gastric cancer of 70 percent in people under the age of 40 and a *sevenfold* increase in the incidence of GEJ gastric CAs. The other reason is the possible presence of CDH-1 in lobular cancer. What is interesting is that the germline pathogenic variations are not much different between normal individuals and the ones affected by lobular cancer (5.2 percent compared to 6.5 percent). There happens, though, to be a ten times more frequent germline mutation in the CDH1 gene, specifically in lobular CA compared to the ductal carcinoma.[36] This variant is commonly seen in gastric carcinoma. This finding confirms our hypothesis that germline mutations are not indicative of risk in lobular breast cancer. The other protein associated with lobular cancer is the ER associated protein MDC1.

There are twenty-eight cancer predisposition genes, including 12 established breast cancer predisposition genes: ATM, BARD1, BRCA1, BRCA2, CDH1, CHEK2, NF1, PALB2, PTEN, RAD51C, RAD51D, and TP53. DNA damage repair genes are recognized in different types of cancer; for example in prostate cancer the main ones recognized are ATM, BRCA1/2, FAM17, GEN1, BRIP1, MLH1, MSH2, MSH6, PALB2, and RAD51C.

The BRCA1 gene usually is associated with basal phenotype and triple negative breast cancer and BRCA2 usually is associated with luminal type (ER+) tumors.

As we all know PARP inhibitors started to be used based on the concept of homologous recombination repair genes. Homologous recombination genes are involved with double stranded DNA repair, and when one is mutated, the DNA depends on single DNA stranded repair mechanisms: nucleotide excision, base excision repairs, and mismatch repair pathways (PARP).

Further, pharma started to sponsor trials and promote using this class of drugs in tumors without HRG. For example, in prostate, ovarian, and breast cancer they were studied regardless of the HRG status. Either way, the overall survival was not changed in many trials using these drugs due to resistance.

When it comes to resistance to PARP inhibitors and platinum agents, which are widely used in patients with BRCA mutations (with both ovarian cancer and triple negative breast cancer), the scientists believe that exon 11 mutation in BRCA confers resistance to platinum, and TP53 mutation confers resistance to PARP inhibitors. That said, the epigenetic influence on such resistance is discounted. In my practice, unless a patient is under epigenetic therapies, to inhibit mechanisms of resistance at BRCA gene (methylation), the patient is advised not to receive any of the other therapies. This is for two reasons. The first is the opening of the BRCA target for the PARP inhibitor to be effective and the second is to reduce the chance of resistance, both through induction of the target by reducing hypoxia and by reduction of the TP53 mutation.

For prostate cancer, besides the BRCA gene, HOXB13, and in pancreatic cancer, CHEK-1 and PALB2 are suggested. Also, ARID1, PTEN, CCNE, and RAD51 are all suggested to be DNA damage repair genes. As such, it is reasonable to consider combining cyclin-dependent kinase inhibitors with PARPi. PTEN loss has also been associated with poor prognosis in EGFR mutated lung cancer.

The standard of care for metastatic chemo naïve prostate cancer is AAP (abiraterone, apalutamide, and prednisone), but none of my patients in this setting receive this blind shot approach, for reasons explained in regards to the activation of basal cells in a metastatic setting.

That said, the pathological variant alone does not determine the cancer risk, as it is influenced by the microenvironment and epigenetics. (And interestingly, the presence of these mutations are not prognostic for survival; for example, in TN breast CA is not dependent on the presence of these mutations in the carriers.[37] That said, other studies are conflicting about this, as the prognosis is worse in TNBC when there is brain metastasis, and is more common in BRCA+ patients. Younger patients obviously have worse prognosis with breast cancer. Increased LDH, over 1.5 to 3 times normal, is also prognostic (average survival is 4 months; 20 percent 1-year survival is reported). That said, in my practice I have had cases where the patient has survived beyond that, depending on the genetics of the tumor.

When discussing germline mutations, we should not forget the epigenetic germline defect. A perfect example is Wilms tumor where, especially in the anaplastic form, the tumor originated from fetal tissue tends to grow very quickly in a child with abdominal mass and a germline mutation of WT1, but these patients, although diagnosed with a genetic syndrome (Beckwith-Wiedemann), only go on to develop a Wilms tumor if the genetic abnormality in chromosome 11p15 is accompanied by a hypermethylation of this gene. This concept is at times mistaken with other somatic events or mutations, as we see in the BRCA gene, when a germline mutation of one SNP only increases the risk of carcinogenesis after there is a somatic loss of the remaining wild-type allele. That said, with the notion that somatic mutation is *not* needed by epimutations, such a concept is revolutionary.

The epigenetic influence of carcinogenesis is driving the tumor incidence in younger individuals with or without germline mutations. In breast CA, germline mutations (protein-truncating variants or missenses) in ATM, BRCA1/2, CHEK2, PALB2 are mostly relevant and TP53, NF-1, BRAD,

RAD (triple negative CA), and PTEN have been discovered to relate with the presence of ER± status of the disease.

Endometrial cancers are of three types genetically: POLE mutant, MSI-high, and TP53 mutant. Approximately 5 percent of endometrial CAs are due to Lynch syndrome, which is a mismatch repair defect, manifesting as a germline *or epigenetic* MMR defect (about 50 percent of hereditary non-polyposis colon CAs have a detectable genetic mutation in mismatch repair genes).

One percent of ovarian cancers are also due to Lynch syndrome. Five tumor suppressor genes have been detected in MMR: MLH1, MSH2, PMS2, EPCAM and MSH6. Patients with Lynch syndrome have a *24 percent to 51 percent* lifetime risk of endometrial cancer compared to 2 percent for the general population, and a *3 percent to 20 percent* lifetime risk for ovarian cancer, compared to 1 percent in the general population. Interestingly, all germline MMR patients had non-serous, non-mucinous ovarian cancers.

Why such a wide gap, from 3 percent to 20 percent increased risk? It is all due to the epigenetic influence of mismatched genes. This is a perfect example of why the germline mutations do not reflect the entire risk of carcinogenesis. Universal tumor testing is now a standard for tumors, but not epigenetic testing (sequential IHC with MLH promoter methylation analysis is what we recommend).

That said, constitutional mismatch repair deficiency syndrome is a lethal cancer predisposition syndrome characterized by early onset synchronous and metachronous multiorgan tumors. These tumors are transformed from benign lesions in the GI and heme and brain 80 percent to 100 percent of the time. Biallelic pathogenic variants lead to MMR deficiency during DNA replication, causing rapid accumulation of mutations (hypermutant phenotype), causing chemo and radiation to be ineffective. The penetrance of this genetic situation is very high (similar to Li-Fraumeni syndrome and hereditary breast-ovarian cancer). For example, a colon polyp becomes malignant in a few years in almost every case, and as such colectomies are recommended for all children with this situation!

As such, surveillance is very important in this group of patients (abdominal ultrasounds, endoscopies, and brain MRIs are recommended every few months), starting around age 4-6 (the median age of diagnosis is 10 years).

To me this is a very impractical screening tool. We must be able to screen these individuals with much more practical and effective tools, and be able to provide meaningful preventive measures beyond imaging.

The table on pages 58–60 is adapted from the Journal of Clinical Oncology August 2021 issue:

Also, the use of immune checkpoint inhibitors in MMR-deficient tumors is questionable. As drugs such as dostarlimab get approved, the higher the cost of care and the less efficacy reported.

In the next chapter we will be discussing the germline mutation screening test and the fact that only one third of patients with ovarian cancer are screened, although all ovarian cancer patients are suggested to be screened. The same is true for breast cancer patients in a metastatic setting; all need to be tested for germline mutations. The presence of breast CA germline variations is about 3–4 percent. Recently, it was suggested that all advanced cancer patients be screened for germline mutations.[38] In the advanced setting of about 50 percent of discovered mutations (9 percent of total checked) have been attributed to BRCA1/2, CHEK2, ATM, and MMR (12 percent); and PALB2 (5 percent).

On the other hand, many screening tools recommended to patients are not scientific. According to a study published in JAMA, the recommendations made for breast cancer prevention based on mammographies are all financially motivated.[39] Also, the effort as far as providing people with the HPV vaccine to reduce the risk of cervical cancer is showing no relationship to the actual subsequent risk.[40]

A new way of screening for ovarian cancer includes Pap smear studies for CA 125, HE4, and mesothelin.

There are only four types of cancer currently being screened for early detection: lung, breast, colon, and cervix (five if you add melanoma), leaving a total of 72 percent of cancers without a screening test to detect them. As far as lung cancer is concerned, a screening low-dose CT scan of the chest is now recommended for patients over the age of 50 with at least 15 pack-years of smoking, including those who have quit in the last 15 years *(See diagram page 61).*

Unfotunately, by the time a tumor is about 2 mm, large enough to be identified in lung cancer screening (the spiral CT threshold), it has already passed the 21-day doubling time. Between the 21- and 28-day doubling times, the tumor increases suddenly to 8 mm, where it can detected by most X-rays.

Classification of Germline Susceptibility With Therapeutic Actionability

OncoKB Level of Evidence[a]	Gene(s)	Drug Class	Examples of Implicated Therapies	Clinical Implications
1	*BRCA1* and *BRCA2*	PARP-1	Olaparib Rucaparib Talazoparib Niraparib	FDA approval: advance ovarian, breast, pancreas, and prostate cancers
1	*RET*	Tyrosine kinase inhibitor	Selpercatinib	FDZ approval (RET-associated)[b]; medullary thyroid cancer
1	*PTCH1*	Hedgehog-signaling inhibitor	Vismodegib	FDA approval (PTCH1-associated): locally advanced basal cell carcinoma
1	*TSC1, TSC2*	mTOR inhibitor	Everolimus	FDA approval (TSC-assobiated): subep-endymal giant cell astrocytoma (SEGA) and renal angiomyolipoma
1	*ALK*	ALK kinase inhibitor	Brigatinib Lorlatinib	FDA approval[b]: NSCLCs with ALK onco-genic alterations
1	*EGFR*	EGFR inhibitor	Osimertinib	FDA approval (EGFR-mutated; gastroin-testinal stromal tumors.
1	*KIT*	Tyrosine kinase inhibitor	Imatinib	FDA approval (cKIT-mutated)[b]: gastroin-testinal stromal tumors

Classification of Germline Susceptibility With Therapeutic Actionability *(cont.)*

OncoKB Level of Evidence[a]	Gene(s)	Drug Class	Examples of Implicated Therapies	Clinical Implications
1	*NF1*	MEK 1/2 inhibitor	Selumetinib	FDA approval (NF1-associated): inoperable plexiform neurofibromas
1	*ATM*	PARP-1	Olaparib	FDA approval: advanced prostate cancer
1	*PALB2*	PARP-1	Olaparib	FDA approval: advanced prostate cancer
1	*RAD51C* and *RAD51D*	PARP-1	Olaparib	FDA approval: advanced prostate cancer
1	*BRIP1*	PARP-1	Olaparib	FDA approval: advanced prostate cancer
1	*BARD1*	PARP-1	Olaparib	FDA approval: advanced prostate cancer
1	*CHEK2*	PARP-1	Olaparib	FDA approval: advanced prostate cancer
1	*RAD51B*	PARP-1	Olaparib	FDA approval: advanced prostate cancer

Classification of Germline Susceptibility With Therapeutic Actionability *(cont.)*

OncoKB Level of Evidence[a]	Gene(s)	Drug Class	Examples of Implicated Therapies	Clinical Implications
1-MSI-H	Lynch syndrome (*MLH1, MSH2, MSH6, PMS2*, and *EPCAM*	Checkpoint inhibitors	Pembrolizumab	FDA approval[b]: advanced MSI-H/dMMR solid tumors (not Lynch syndrome specifically)
3B	All level 1 genes	See level 1	See level 1	Drug class has FDS approval (level1) in a different cancer type
3B	*MET*	MET kinase inhibitor	Cabozantinib	Renal cell carcinoma
4	*POLE* and *POLD1*	Checkpoint inhibitors	Durvalumab	Under clinical investigation
4	*CDKN2A* and *CDK4*	CDK4/6 inhibition	Abemaciclib Palbociclib Ribociclib	Under clinical investigation
4	*SDHB*	Alkylating agent	Temozolomide	Under clinical investigation
4	*FH*	EGFR inhibitor and anti-VEGF	Erlotinib Bevacizumab	Under clinical investigation
4	*VHL*	HIF-2α	MK-6482	Under clinical investigation

Abbreviation: dMMR, defective DNA mismatch repair; FDA, US Food and Drug Administration; MSI-H, high-frequency microsatellite instability; NSCLC, non-small-cell lung cancer; PARP-1, poly(ADP-ribose) polymerase inhibitor.

[a]Further data supporting the current OncoKB level of evidence can be found at OncoKB website.

[b]Designates that FDA approval is not specific to the germline alteration.

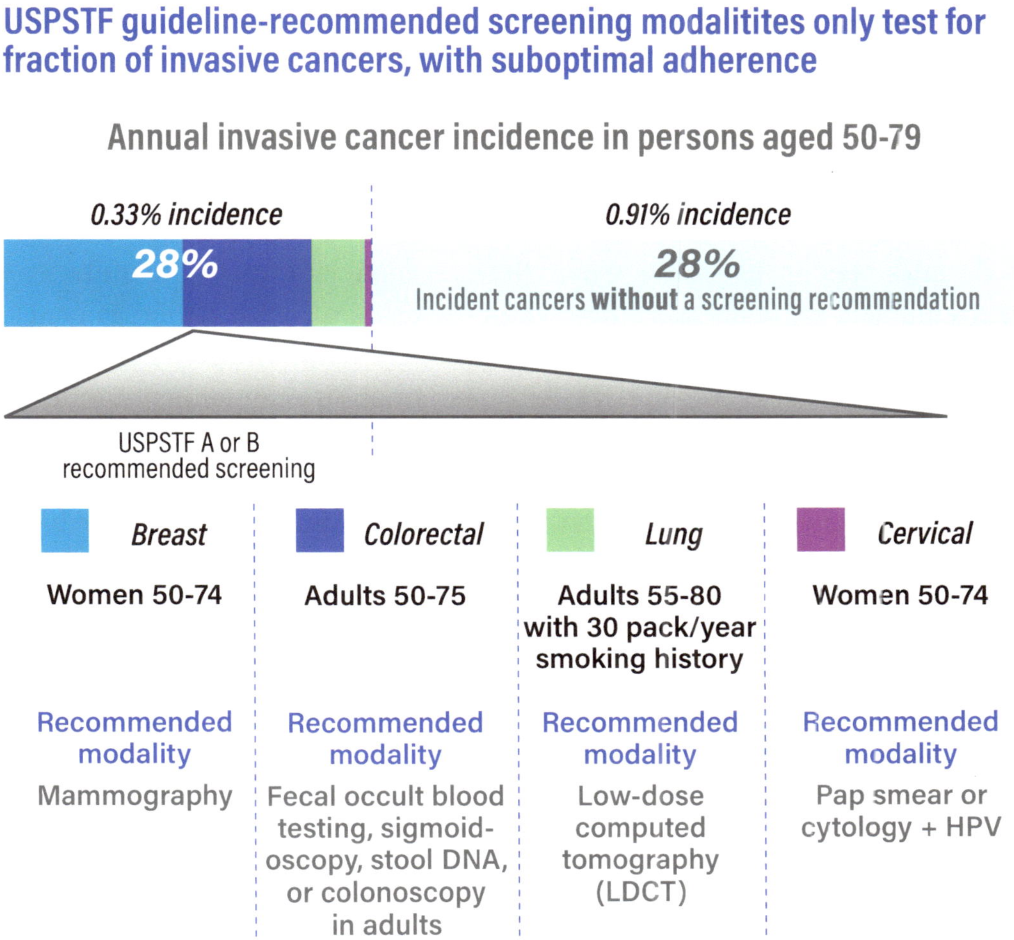

This recommendation was changed in 2021 from 30 pack-years of cigarettes to 20, but as we will discuss in more detail, we now know that the risk of lung cancer depends not on the smoking exposure of the individual him- or herself but on the exposure of the prior generations, as frequently, nonsmokers are diagnosed with lung cancer (specifically with KRAS mutations), which are a hallmark of smoking exposure, secondary to the epigenetic disruption that occurred in prior generations. Understanding this mechanism makes one wonder how the standard recommendation of a low-dose CT scan could make any logical sense, as it only counts that single person's exposure to smoking...while the damage has happened in prior generations by the inhibition of histone demethylases.

Finally, I would like to reemphasize that the genomic instability of a tissue does not automatically transfer to carcinogenicity. For example, normal placenta tissue has the largest mutations and in fact a patchwork of tumors, with chromosomal abnormalities from 11q15 to SNPs. Yet the placenta does not detect aberrant cells and delete them and the baby is completely normal and genetically separated from the placenta. A good example of a fetal originated tumor is neuroblastoma, where the epigenetic control of the mutations is dysregulated. Only about 16 percent of the cases were identified as genetically caused.[41] Genetic mutations in neuroblastoma include ALK, MyCN, CHEK2, BARD1, NSD1, and RMRP (naxitamab, a monoclonal antibody, was recently approved for neuroblastoma). Current research has identified poor prognostic markers such as FOXR1 and c-Myc in neuroblastoma, which are epigenetically controlled. Conventional therapies recently have included cyclin-dependent kinases in combination with temozolomide. For ALK ensartinib has shown better results than crizotinib, with better CNS penetration. I personally have never used crizotinib in patients with CNS disease, as I have been very doubtful about its penetration to the brain. My choice in this circumstance has been ceritinib.

Chapter Six

Prevention First

In general, the US is worse in access to health care and preventive measures when compared to almost every other comparable country.

Mortality amenable to health care is a measure of the rates of death considered preventable by timely and effective care. While the health care

The U.S. has higher rates of medical, medication, and lab errors than comparable countries

The U.S. ranks last in a measure of health care access and quality, indicating higher rates of amenable mortality than peer countries

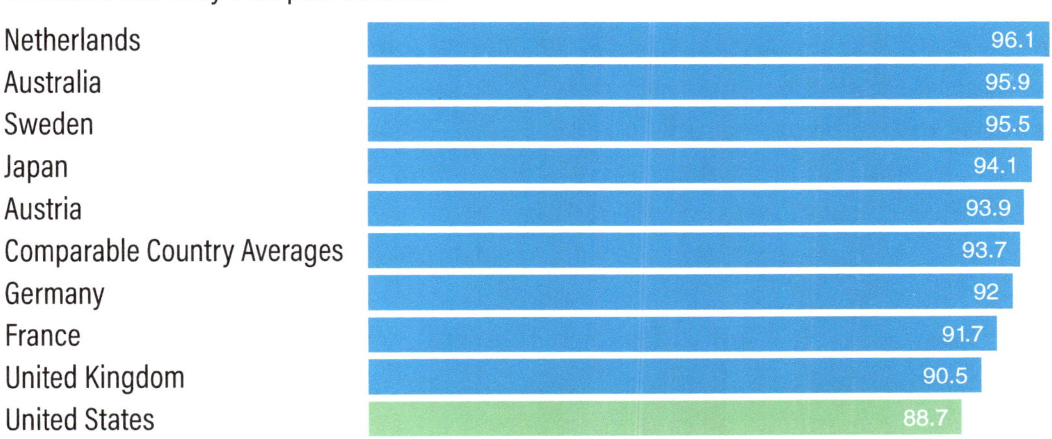

▪ Source: KFF analysis of data from: "Measuring performance on the Healthcare Access and Quality Indes for 195 countries and territories and selected subnational locations: a systematic analysis from the Global Burden of Disease Study 2016' The Lancet, 23 May 2018

system might not be expected to prevent death in all instances, differences in amendable mortality indicate how effectively health care is delivered. The Healthcare Access and Quality Index is scaled from 0 (worst) to 100 (best) and is based on amenable mortality. The HAQ Index uses age-standardized, risk-standardized mortality rates for 32 causes of death that timely and effective health care could potentially prevent. Lower scores indicate high mortality rates for causes amenable to health care, while higher scores indicate lower mortality rates, possibly reflecting better quality and access to care. The US ranks last among comparable countries on the HAQ Index, with a score of 88.7.

In this section I would like to discuss the role of preventive oncology and the fact that the preventive methods and tools available to mainstream medicine are extremely deficient. Although traditionally the American Society of Preventive Oncology has been responsible for this task, the focus on preventing invasive cancer has been the focus of many studies. Reformatting of the current model of prevention is necessary to address the current unmet need. The common knowledge of oncologists when it comes to primary prevention, as well as secondary prevention, is lacking. The simple question about whether breast implants cause cancer is still answered with "we do not know," when studies have certainly shown that to be the case.

Recommendations for prevention for breast, as well as prostate, cancer in men are simply insufficient. We know that men with BRCA mutations have a 20–30 percent chance of prostate cancer, compared to 1–2 percent for breast cancer. Neither prophylactic prostatectomy or mastectomy is recommended for them. That said, survival depends mostly on early detection of patients' disease. For example, an individual with high-risk prostate cancer (with a Gleason score of more than 8, PSA > 20, and grade 3) has less than a 25 percent chance of relapse-free disease, compared to early detection with a 75 percent cure rate. Overall, 50 percent of patients who are diagnosed with prostate cancer are treated with active surveillance and amongst them, 50 percent over time will be switched to active therapy (in 3 years).

The routine physical examination to find pre-melanoma lesions on the skin is something I have always thought to be senseless. What if someone could not have an annual examination of their entire skin, or the physician examining them did not see a pre-cancerous lesion? Then what? And why should we wait for the precancerous lesion to first appear before we start to think how to stop the process, and even then with a meaningless procedure like a shave or a core biopsy, which can itself disseminate the tumor cells? As we know, more people die of early thin melanoma (T1) disease (less than 1 mm in Breslow thickness) than more advanced disease, as the incidence is higher with T1, so what if we can prevent T1 from occurring by examining the risk of patients by detecting the genomic risk?

The early detection concept of using breath analysis was studied by researchers at the University of Latvia, using the same concept that dogs can smell cancer by breathing near the affected individual (volatile organic compounds).

A recently studied biomarker is Profilin 1 protein, which is involved in cellular activities, motility, survival, and membrane trafficking, and attributed not only to cancer, but also other conditions such as ALS and hypertension. It can be used as a biomarker for diagnosis of cancer.

For bladder cancer, different assays are now available, from Cxbladder (looking at 5 m RNAs such as IGFBP5, HOXA12, CXCR2, and CDK1), Assume MDx (looking at HRAS, FGFR3, and TERT, plus methylation signatures for other genes), UroMark and Urobest, and Bladder EpiCheck (looking at different methylation markers), with which a clinician can detect a noninvasive bladder cancer long before the hematuria is detected (70 percent of bladder cancers are found in the early stage).

We also know that neurotensin is a marker we can detect in blood for cancer risk (with a specificity of 90 percent). It is a marker for obesity, and it is widely associated with increased cancer growth, as cancer cells carry neurotensin receptors at a high level (it is also used as a diagnostic tool in detecting imaging), and is associated with reduced survival, as it increases IL-8. There are significant studies on using NTSR1 antagonist/radioactive drugs to target the neurotensin in different types of malignancies. Obesity also increases the immune suppression. Drugs such as fenfluramine and sibutramine have been banned by the FDA.

The concept of pre-conceptional medicine addresses the fact that mutations start to accumulate at very early gestational ages. For example, in the case of myelofibrosis/essential thrombocytosis, the JAK2 V617F acquisition occurs between 6.2 gestational weeks and 1.3 years of life. The same is true for clonal expansion of driver genetic mutations for DNMT3A, which occurs within days after conception. Clonal expansion rate depends on the clone itself and can range from small rate to high rate depending the stroma. That said, the epigenetic profiling of hematological cancer has been shown to be superior to the genetic markup; for example, in CLL the IGHV1 mutation is inferior to the epigenetic classification of the disease.[42]

Integration of multiomic data with graph convolutional networks to identify new cancer genes and their associated molecular mechanisms is a new algorithm that predicts which gene causes cancer, even if the DNA sequence is normal! These are well-known genes that "influence" the network of genetics through epigenetic mechanisms.

Also, MethSig is a new algorithmic detection that predicts driver methylation changes, which is not tumor type specific (methylation drivers happen in dozens, compared to thousands of passenger methylation events).

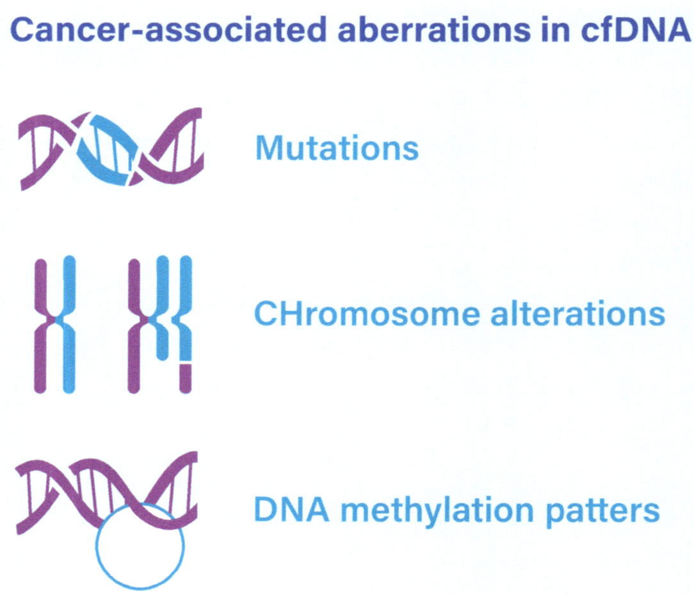

Early detection of urothelial cancer can be performed with Urine Exfoliated Cells Copy Number Aberrant Detector with 90 percent specificity, where urine cytology has false negative results 35 percent to 50 percent of the time. Early detection of ovarian cancer can be performed using carbon nanotubes (nano sensors) with AI on serum biomarkers of normal individuals.[43] Also, using acid-sensitive nanoparticles to accumulate copper 64 in the tumor microenvironment provides an imaging technique to detect tumors. These nanoparticles are very vital in diagnosis of tumors.

Screening other types of cancer, such as colon cancer, has been modified to younger ages, 45 years for colon CA. The reason is that since 2015, the rate of incidence has increased from 7 percent to 30 percent in people aged less than 40 years, by 8 percent per year in people aged 20–29, and by 5 percent per year in people aged 30–39. Although this has been blamed on germline mutations being more common in younger individuals, it does not explain the high incidence. For example, it is estimated that microsatellite instability is found in about 10–15 percent of the population (regardless of germline mutations) and about 25–30 percent of patients over the age of 50. Lynch syndrome contributes to about 30 percent of these cases. That said, the increased incidence is not fully explained by this picture until we have recognized and added the role of DNA methylation defects (specifically in genes such as LINE-1). We will discuss this phenomenon more in detail in further sections. It is worthy of notice that RAS mutations are not more common in younger individuals or in families affected by germline mutations.[44]

Early detection of lung nodules through artificial intelligence has improved lung cancer screening. This is also true in the DREAM Challenge studies of PHF19 as an epigenetic marker for invasive prostate cancer. This marker in the pathology of prostate cancer is indicative of aggressive disease. PSA screening is now again recommended for men between the ages of 55 and 69. For breast cancer, performing mammographies for patients less than 30 years old is not indicated. Especially when BRCA is positive, MRI is preferred for patients between 25 and 30 (MRI is indicated when the risk of recurrence is over 20 percent). There are studies suggesting serum screening for five proteins for prostate cancer screening: PSA, hK2, MSMB, and MIC-1, along with biopsy to increase its sensitivity. In my clinic, the urine PCA3 gene is the most predictive (with the highest sensitivity), and can replace the biopsy. The urine needs to be collected after a good prostate massage in a rectal examination, about which some of my patients are hesitant, but surely it is a safer approach than biopsy.

Ovarian cancer screening is also not recommended for BRCA+ individuals (as it does not change the survival). Removing the ovaries is not indicated, especially with the risk of HRT in BRCA+ patients. Mastectomy, however, is indicated, as there is a 60 percent chance of contralateral breast CA. Taking AI also seems to be prophylactic in breast CA incidence. Taking OCP reduces the ovarian cancer risk by 60 percent in carriers of BRCA2.

For another example, in a study population of 144,524 women with DCIS, although all the patients diagnosed received surgery and almost half also received radiotherapy, the risk of dying of breast cancer was increased by three quarters after the diagnosis of DCIS. This suggests that the current treatment focused on preventing invasive recurrence is extremely deficient. The majority of oncologists are not fully aware of the process involved in the transition of a precancerous lesion to a malignant tumor. For example, in the case of DCIS, DCIS is itself a cancer with metastatic potential and not a precursor for invasive disease. The identification of DCIS scoring can assist the conventional provider in assessing the risk benefit of using radiation (5 percent absolute risk reduction in low-risk cases), but if we believe that a 10-year risk reduction assessment is predictable by a 16-gene assessment, we simply have underestimated the role of epigenetic influence.

In fact, some of my patients with a diagnosis of DCIS have had positive circulating tumor cells for breast cancer, meaning that although the pathology did not identify any malignant invasive ductal carcinoma the cancer cells were seen and quantified in a blood sample. This means that these patients not only had breast cancer, they also had micrometastatic stage IV breast cancer. It is no surprise that after surgery and radiation, which is the standard of care for the majority of these DCIS patients, the disease manifests itself more significantly in the form of metastatic disease. This concept of minimizing surgical insult is becoming more and more a focus for new studies; for example, in ovarian cancer, minimally invasive cytoreduction is becoming the preferred surgical intervention.

Another example is early stage noninvasive bladder cancer that can be treated with intravesical BCG to prevent disease progression, although 40 percent of these cases recur, and cancer vaccines can be used to prevent progression. An example is an adenovirus delivering IFN alpha-2b gene (nadofaragene) to the bladder wall. Traditionally we use valrubicin. In advanced bladder cancer (extended to the myometrium) we can use either cystectomy or combo immune therapy after the traditional four-cycle chemotherapy. In my practice the use of epigenetic therapies has shown very good results in patients who refuse cystectomy or immune therapies in this case.

We all know that by inducing stem cell activation both surgery and radiation can increase the potential of metastatic disease when the disease is not already localized. The presence of Wnt activation enhances the EMT process. A good example of this is medulloblastoma, where now the studies suggest lower doses of radiation are better![45] The Wnt pathway is activated in 100 percent of cases of pediatric medulloblastomas. Isotretinoin as a proapoptotic drug can add to patients' response through this target.[46] The addition of isotretinoin has added a 20 percent improved survival rate in neuroblastomas.

In my practice, the application of radiation is limited to palliative uses, at low dose. Studies have shown that a dose of 8 Gy in a single fraction is considered the gold standard for skeletal metastatic disease. Stereotactic body radiation with application of higher doses of radiotherapy (24 Gy) in two sessions are also acceptable, in my view, but only if there is no exposure to the gut. Cases of ischemic colitis have been reported as a result of nontargeted radiation of the spine in elderly individuals.

For example, systemic aortic and pelvic lymphadenectomy does not improve survival in ovarian cancer, compared with simple node sampling. Also, adjuvant chemotherapy (chemotherapy given 4–6 weeks post surgery) has no role in improving survival in mucinous, grade 1 endometrioid, and low-grade serous ovarian cancers), compared to observation alone. Nor do angiogenesis inhibitors have any role in stage I ovarian cancers, clear cell tumors, or tumors harboring BRCA mutations. Using chemotherapy beyond six cycles (or three cycles in stage I high-grade serous), or triplet chemo agents has also failed to improve survival despite significant additive toxicities. There are a few new agents with marginal efficacy in ovarian cancer, including rebastinib (a TIE2 inhibitor) and mirvetuximab (a folic acid receptor antibody conjugate) in combination with platinum agents. The argument again would be that the folic acid receptor activation would enhance the tumor division by mitosis.

That said, there are studies showing that modulated hyperthermia is used for cervical cancer along with traditional radiation therapies (stages II and III) EBRT, with impressive local disease control for 2 years. Tisotumab has been approved recently for cervical cancer after a first line of Avastin plus chemotherapy (response rate 10 percent).

We also know that chemotherapy increases the risk of CNS metastasis in prostate cancer, as studies have shown that.[47]

Nowadays they are also recommending low-dose radiation for Covid-19 pneumonia. Can you believe this?? Another study confirmed the appearance of axillary LN in 12 percent of vaccinated individuals and recommended mammograms to rule out malignancy! (Lindsey Nolen, Oncology Times, April 5, 2021, Volume 43, Number 7, pages 16–17).[48] Again, the National Comprehensive Cancer Network issued a recommendation for vaccination for all cancer patients, "understanding that there are limited safety and efficacy data in these patients"![49]

I also want to emphasize the role of genetic testing on patients who have a high risk because of their family risk factors. There is not a commercially available genetic or epigenetic testing that can identify those at risk, except limited genetic mutation testing such as BRCA or My Risk Panel, while at this time some laboratories offer a 25 to 28 genetic panel. It is very important to know that the mutation of a gene does not rule out the epigenetic appearance of that gene.

This means that if a gene is not mutated it still can be completely dysfunctional and as such lack of mutation does not rule out malfunction of a gene. Yet, as we speak, we unfortunately do not have commercially available testing that can identify epigenetic appearances that are equivalent to mutations. If we can identify the patients at risk long before the cancer presents itself, it can easily be prevented by augmentation of the non-mutated genes that are in a complex where they crosstalk with each other. For example, in the case of the BRCA gene there is a superior gene called PTEN, so that when the BRCA is mutated or for that matter methylated, by augmentation of the PTEN gene, you can recover the function of the deficient BRCA and significantly reduce the carcinogenesis in that patient.

Here I'm going back to the study of DCIS, which was a very large study between the years of 1995 and 2014. Although all the women had surgery, and almost half of them had radiotherapy, 45 percent of these patients who had either an invasive recurrence on the same side or contralateral invasive breast cancer in the interim between DCIS and breast cancer died. In other words, 27 percent of these patients had mastectomies, and although the mortality rate for death from breast cancer related to DCIS was only 3–4 percent, this study recommended chemotherapy for younger individuals and Black women.

The death rate from metastatic breast cancer in three months *from diagnosis to death* is 16 percent. One out of six patients die before the three-month mark following diagnosis. HOBX2 hypomethylation is found to be most common in luminal tumors and associated with a worse prognosis. ATOH8 hypomethylation is mainly seen in triple negative breast cancer. As such, these biomarkers have been found to be the most relevant epigenetic marks in breast cancer to associate with therapy response and outcome. It is noteworthy that the JAK/STAT and SWI/SNF pathways are dysregulated at higher rates in metastatic disease, or Bcl-2 overexpression is common in this setting, yet drugs such as venetoclax have failed to show improved outcomes.

This is to note that, though breast cancer has not been diagnosed, their thought is to start chemotherapy because it is hypothesized that a patient with DCIS will ultimately develop breast cancer. That said, there is no study that proves that using chemotherapy on younger individuals or Black women has any potential in improving survival. That's another big issue—that until physicians and researchers start looking for what will prevent the disease from progressing from DCIS to actual cancer, everyone is missing the whole point that the role of epigenetics is the real key player in this transition from precancerous lesion to extensive disease. Yet, chemotherapy can do absolutely nothing to reduce the risk.

The fact of the matter is that even after 25 years of early diagnosis of DCIS or breast cancer patients are still at risk for recurrence. No one has asked the question why, after local disease control and surgery, the patient is still at risk for cancer even after 15 years. The simple answer to this question is that the cause of recurrence is the stem cell activities because of epigenetic variances. As such, you can say that patients who have high risk of pre-lesions or pre-cancerous lesions pretty much have the same risk of death as if they had an actual disease. For example, in the case of breast cancer the risk of death is about 3–4 percent, which is similar to the 5 percent recorded for small-size breast cancer, and as such you can conclude that patients with high risk will at one point develop cancer, and it's just a matter of when. Ultimately, the patient's demise is independent of the precancer treatment.

Understanding the role of stem cells is essential because otherwise a normal practitioner with no such knowledge would think that DCIS has a risk of only 15 percent for development of invasive local recurrence. But interestingly, preventing the invasive local recurrence has nothing to do with preventing death, because the patient's recurrence is independent of the presence of DCIS. As such, it is independent of its treatment, which in turn is merely dependent on the cause of the DCIS, which is stem cell activity.

There have been significant studies relating different types of tumors to viral infections. For example, the role of EBV in head and neck cancer has been shown. It even is suggested that after patients complete chemo and radiation, the need to continue chemotherapy depends on the presence of *Plasmodium falciparum* EBV DNA in the plasma.[50]

The note that epigenetic regulation caused by microbiomes/bacteria is essentially driving tumorigenesis is very interesting. One example is the microbiome in the breast that plays a significant role in the carcinogenesis of breast cancer. Another example is Firmicutes and Cyanobacterias that are present in 100 percent of endometrial tumors, and veilonella and lung tumors (with increased IL-17). It is now evident that when one allele of the TP53 gene is altered, the activity of the gene becomes impaired (dominant negative effect), and then the concomitant loss of the wild-type TP53 allele with the presence of the mutated one leads to the aberrant production of mutant TP53 proteins, which gains oncogenic activity (gain of function fosters the oncogenic activity), then the gut microbiome can switch the activity of the mutant TP53 protein from tumor suppressive to oncogenic activity. This mechanism is completely under the influence of epigenetic control. A study shows that cancer mutations are *plastic* and only penetrate when the microenvironment is in favor of that.[51] Another finding is that gallic acid, which is mainly in the leaves of natural food, causes exactly the same effect (activation of oncopromoter TP53). One of these bacterias is *Bacteroid fragilis* that generally grows in individuals suffering from dysbiosis after being exposed to penicillin in the right side of the colon. The magnitude of effect on carcinogenesis seems to be related to decreased production of butyric acid in the gut,[52] the same epigenetic concept discussed by Dr. Bruzynski decades ago. That said, we have to understand that there is also a known paradoxical effect of butyrate inducing proliferation of harvested MSH-2 deficient colon cells. As such, I would recommend that patients with DDR refrain from butyric acid containing oral therapies. That kept in mind, butyric acid plays a significant role as part of my standard protocol to prevent and therapy of DDR deficient tumors.

Other recent studies have shown that *Fusobacterium nucleatum* can be a potential biomarker in the stool for CRC (it activates the Wnt, beta-catenin signaling, and toll-like receptor 4 pathway down to NF-kappa B. As such this microorganism is suggested to confer resistance to standard chemotherapies used for colorectal CA (platinum and 5-FU).

Commonly found *E. coli*, and rarely found *C. difficile*, can produce the genotoxin colibactin to cause DNA damage and crosslinks as well as chromosomal abnormalities. *Bacteroid fragilis* can produce similar toxins, activating IL-17 and STAT3, E-cadherin, and c-Myc. These tumors show DNA hypermethylation (of oncosuppressor genes such as RUNX1) compared to the spontaneous tumors in APC. *Enterobacter faecalis* and *peptostreptococcuses* produce reactive oxygen species in the microenvironment of the colon. Also, blood infections at times are associated with CRC (*Streptococcus gallolyticus*). Periodontitis generates oral microbe-reactive T helper cells that are imprinted with gut tropism and migrate to the inflamed gut, resulting in activation by translocated oral microbiota and subsequent development of colitis.

It has been suggested that oral antibiotics increase the risk of colon cancer and reduce the risk of rectal cancer (Volume 19 (2021): Issue 8 (*Aug 2021*), Journal of the National Comprehensive Cancer Network. Online ISSN: 1540-1413).[53]

On the other hand, the microbiomes generally known as protective (and correlated with improved survival) are *Streptomyces*, *Pseudoxanthomonas*, *Bacillus clausii*, and *Saccharopolysporas* (American Association for Cancer Research (AACR) Virtual Conference 2021),[54] as well as *Geobacillus*, *Burkholderiales*, and *Pseudomonas* (Developmental Therapeutics 2021).[55] In breast cancer, *Gemmiger formicilis* caused a better response to therapy and *Bacteroides* is associated with lower response.

Yes-associated protein 1 (YAP-1) counteracts the TP53. So the lower the YAP-1, the higher the TP53 mutation, and another mechanism of resistance to immune therapies. The presence of microRNA-130b correlates with targeting TAp63, which is a functional overlap with TP53. Retinoic acid, for example, activates TAp63. Bcl-2 inhibitors indirectly target the downstream pathway of TAp63 (mostly in ovarian CA and lymphoma).

Therefore, preventing tumors by early detection of viral infection and treatments to suppress the viral load is essential. Interestingly, some clinicians and researchers are *against* early detection, as they feel it postpones the actual therapy![56] I sometimes wonder if we live in a Third World country with no access to technology....

It is also well documented that barium enemas, GI series, and CT exposures of the abdomen and pelvis significantly increase the risk of testicular cancer.

Also as far as secondary prevention techniques, the use of new technologies to screen patients (like PSMA scans in prostate cancer) has helped to detect early recurrence after definitive treatments. That said, the standard techniques are insufficient in detecting microscopic disease in so-called remission time. The doubling time for the PSA if less than 10 months justifies using androgen receptor inhibitors, such as enzalutamide (at a low dose of 80 mg a day), darolutamide, or apalutamide even when the scan is negative (PSMA is usually positive in this setting). The side effects with these agents are a major concern (although less with darolutamide), and include falls, fractures, and seizures. The sooner AR receptor inhibitors are started the better in OS, after definitive therapy in a nonmetastatic castration setting. The choice for treating metastatic castration resistant prostate cancer is between docetaxel and enzalutamide (or radium 223). It is of note that using enzalutamide *after* docetaxel has not shown any improvement in survival.[57]

The combination of radium 223 and enzalutamide has not shown any improved survival though PSMA-targeted radiotherapy has improved survival for 6 months in patients with metastatic prostate cancer (Lu-PSMA 617). That said, radium has increased the risk of fracture in these patients.

The other strategy is to apply androgen therapy in a bipolar fashion (to reduce the AR receptor up-regulation). Studies have shown that this method has similar results to standard androgen therapy, with fewer side effects, and should be considered for clinical integration.[58]

This finding is in line with the concept of prostate cancer treatment at our clinic, where patients do *not* follow standard androgen deprivation therapy (ADT) regimens and protocols. Traditionally GnRH antagonists and LHRH agonists are used as the ADT, Lupron as an injectable form and relugolix as an oral form of GnRH antagonist (degarelex is also an injectable form). The main difference is in the incidences of cardiovascular disease.

Recent research has shown that using any androgen inhibition in patients with prostate cancer not only increases the risk of bone metastasis but also increases the gut overgrowth of the microbiome that causes androgen resistance (*Ruminococcus* vs. *Prevotella* colonies).[59]

It is also worth mentioning here that the standard of care, which is a combination of chemotherapy with hormonal blockade, has been shown to have similar results to ADT alone, with higher toxicity.[60] This is after decades of oncologists' arguments and millions of patients being exposed to the chemotherapy as the first-line treatment in combination with ADT.

Androgen inhibition has also been suggested in some other types of cancer, such as salivary gland tumors. Obviously in breast cancer, about 70–95 percent of them are AR+ (a very surprisingly high number, compared to the original thought of 60 percent only in TNBCA), presented by Erica Mayer in *Oncology Times* in November 2021.[61] Enobosarm has been suggested if the staining cutoff is reached in about 40 percent of cases.

Our experience has consistently shown that metastatic disease is produced by CTCs, which do not carry AR at all. We also have shown that the basal cells responsible for EMT transition are upregulated due to the selective advantage created by ADT. The work of Owen Witte at University of California, Irvine, clearly showed that the responsible cells for patients' advanced metastatic disease are basal cells differentiated by the presence of mutated AR, FOXA1, TP53, RB1, and c-Myc. These cells manifest *small cell* neuroendocrine prostate cells. We also know that this Darwinian mechanism of resistance is controlled by epigenetics through the interaction of NTD1 and the LBD portion of the receptor, and overexpression of AR-V7 (there are studies suggesting BET inhibitors and RV-7 inhibitors in this setting). BET inhibitors are also being studied in treatment of myelofibrosis, along with JAK inhibitors.

My personal interactions with Owen Witte, the scientist who discovered Gleevec, were very informative. He gave an amazing talk about prostate cancer at UCI in 2019, where I was also in attendance to present my research (Owen Witte, MD is the founding director of the UCLA Broad Stem Cell Research Center). That said, he was unable to answer a question I had about prostate cancer basal cells, and that was in relation to their AR. Long story short, it took mainstream medicine a very long time to realize that testosterone inhibition increases the tendency of the prostate cancer cells to metastasize to the bones. A recent study looking at 819 patients with intermediate risk (PSA less than 20 and more than 10, and Gleason score of greater than 7) showed that overall survival of patients receiving androgen deprivation compared to patients who only receive radiation is the same after 10 years.[62]

In patients with germline mutations, strict screening tests are indicated, for example with Li Fraunani syndrome, annual screenings, MRIs, whole-body ultrasounds, labs, and so on are suggested. Mammographies are indicated, but knowing the high risk of radiation in patients with DNA damage repair gene defects, I certainly do not advise them. Currently, digital breast tomosynthesis, studied in Sweden, is suggested to replace the standard digital mammograms; my preference remains with breast MRI. Breast thermography, which has been offered by many practitioners (my clinic offered it for a decade), can be used in conjunction with other screening tools, and mainly can help with information about the vasculogenesis around the tumor.

Secondary Prevention

It is a surprising but real fact that the standard of care for secondary prevention is not providing any meaningful plan for prevention. First of all, as soon as a patient accomplishes remission (which most of the time is not a durable one) every therapeutic effort stops! Second, the patient is not directed toward any proactive measures except surveillance through imaging, which is after the fact. Many studies also have shown that these measures do *not* correlate with patient survival. In other words, detecting the recurrence by imaging or even labs (tumor markers) does not mean that patient will have a better outcome. The reason is obvious: Any therapy that is suggested and provided after a tumor recurrence fails anyway. For example, it was published in ESMO Virtual Congress, in November 2020, "No survival benefit was observed with second surgeries compared with intensive monitoring" for colorectal cancer,[63] where after follow-up for 6.5 years, 22 percent of patients had recurred, mostly in a metastatic setting (75 percent). Despite differences in intensity of monitoring, no differential overall survival benefit seemed to emerge between the arms according to the type of surveillance (i.e., CEA and CT scan surveillance, or neither). Recurrence-free survival was also similar, with no subgroup differences.

Real effective secondary prevention stems from the inhibition of stem cells. The plasticity of stem cells is mainly increased after administration of adjuvant therapies, meaning that as soon as the patient accomplishes so-called remission, the chance for recurrence of the disease increases. The real discordance between the PFS and OS stems from the same reality. It will only be a matter of time before the tumor starts to recur in a metastatic setting, and the patient who was so far celebrating a possible cure, with no findings in the scans, now is called terminal with no viable options.

Although the patient now believes this was no fault of the therapy and everyone is blaming cancer for it, the real truth is that the cancer at times would have only been able to spread after it was triggered by the therapy. This concept was evidently proven studying the effects of alkylators on intratumoral hypoxia (Cytoxan was studied as an agent, as well as Sutent). Interestingly, the use of Sutent in metastatic pancreatic cancer was suggested to be effective when combined with the prodrug evofosfamide, in fact for the very same reason, as evofosfamide becomes active in the hypoxic environment created by Sutent!

We also discussed the activation of the Wnt pathway by radiation and further the role of other targeted drugs such as angiogenic inhibitors in a similar fashion, where they increase the activity of HIF by binding to VEGF (the effects of VEGF are not dependent on its presence, but rather the affinity of its binding to its receptor, influenced by MMP, which is increased by chemotherapy through increased formation of interleukins). One of the known treatments for increasing HIF is HIF prolyl hydroxylase inhibitors, used for anemia in CKD. Chemotherapy does exactly that!

There is only a 7 percent response rate in large B-cell lymphomas to salvage chemotherapy, and patients with refractory disease live on average for only about six months, but we did not know this truth, until the FDA approved the newer drugs. Why? Because we did not want to shine light on facts that related to the "failure of standard of care." Now we all talk about failure of chemotherapies, because we want newer drugs to capture the market. Drugs such as loncastuximab (a conjugate of alkylating agent and CD-19 antibodies) for relapsed or refractory large B-cell lymphomas. The problem with all bispecifics is similar to those with CAR-T therapies, causing cytokine storms (reported in 50 percent of cases!) and on-off target syndrome.

This is also true in other types of cancer, including lung cancer. As we know, only about 43 percent of stage I lung cancer is curable and 57 percent recurs. The simple reason is again something no oncologist has ever questioned. The reality is that the treatments can stimulate the stem cells to produce new cells and cause recurrence. In fact, many times, the CTC if checked is positive from the beginning and these tumors are not staged correctly. A recent study published in *The Oncologist* in September 2021 suggested that in RCC any CTC above 3 or 4 carries a worse prognosis if positive at the base line.[64] A perfect example is a patient of mine treated with only epigenetic therapies when diagnosed with localized RCC. Initial CTC with G250 mutation carries the highest risk of RCC aggressive disease. As such, epigenetic therapies were able to completely eradicate the CTC and further accomplish complete remission in this patient, without use of standard therapies. Other immune therapies for advanced RCC include LAG-3 inhibitors and IL-27 directed antibodies and macrophage checkpoint inhibitors.

In our practice secondary prevention is an essential part of the approach to cancer patients when they accomplish remission, and it is accomplished by reducing the stemness of the cancer stem cells through different markers, including c-Myc. The genetic predisposition for cancer recurrence is controlled through epigenetic regulation even in the most predictable cases, such as BRCA germline mutations.

Chapter Seven

Clinical Trial Design, Next Generation Trials, and Off-label Therapies

HERE I WOULD LIKE TO EMPHASIZE SOME KEY CHALLENGES IN REGARD to the design of clinical trials. The first challenge has to do with tumor type of specific trials. As we all know, the concept of oncology care began with mechlorethamine, which was found in mustard gas used in World War II and discovered as a treatment for leukemia. This was the initial step in designing treatments and drugs based on different types of cancer. This continued until recently, when we realized that basket trials looking at specific genetic biomarkers in each tumor are superior to the standard clinical trial design where the tumor types are treated merely based on the location of the tumor. For example, if a drug is approved for lung cancer the very first indication for investigational new drug application is lung cancer. You can use that drug as off-label for other indications if the merits show that using the off-label will be superior to the on labeled drug, that the on labeled drug is too toxic, or that the on labeled drug has failed.

Some of these off-label drugs later are classified as on-label. For example, arsenic is now approved as on-label for treatment of acute promyelocytic leukemia (along with ATRA and gemtuzumab). Recent studies, especially in the pediatric population, have confirmed that such an approach, if taken collectively, can improve patient outcome. It is my suggestion to improve the patient's outcome by collectively analyzing case studies where off-label targeted drugs are implemented, as indicated by genomically targeted single patient use applications in different institutions. One study has shown an approximately 39 percent response in a pediatric population treated with such an approach, when the standard of care was exhausted and there was no clinical trial available.[65]

That said, at times both the type and the molecular profiling of a tumor would need to be considered. For example, in the case of gastric CA, the presence of HER2 target does not simply confer response to a HER2 therapy, as it does in the case of breast cancer (Herceptin has failed to show improved survival in these types of cancer, as a continuation therapy). Also, interim analysis (based on secondary criteria meeting two thirds of the outcome) can be widely used to reduce the trial duration. One of the main reasons for a patient to develop resistance to anti-HER2 therapy is simply the activation of c-Met. Genistein, a soy isoflavone, has been shown to inhibit this target (c-Met). MicroRNA 34 can be measured to identify this expression and reduced post exposure to genistein. Capmatinib and tepotinib were recently approved for treatment of c-Met+ NSCLC, since in NSCLC, the EGFR mutation and subsequent resistance to therapy is correlated with c-Met activation (20 percent), particularly when osimertinib is used, combination inhibition with selumetinib, (selumetinib is also an inhibitor of MEK and approved for neurofibromatosis), savolitinib, or bispecific antibodies, such as amivantamab, are promoted. C-met activation is accompanied by HGF and downstream activations of PIK3CA and MAPK. MET exon 14 mutations are also particularly enriched in sarcomatoid subtypes of NSCLC. Another medication promoted for use after failure of osimertinib is lazertinib (which is generally used along with amivantamab) in T790M mutations.

Regarding C-met, my experience is that almost all papillary thyroid cancers (or papillary RCC) are C-met mutated, positive, or the equivalent (overexpression, chromosome 7 gain, HGF amplification), and using crizotonib is a good choice (newer MET inhibitors such as savolitinib have shown only a 7 percent response rate). In general, 50 percent of thyroid CAs are also BRAF+, and as such using a combination of BRAF and MEK inhibitors are now approved for this setting (which again I am not optimistic about). Ficlatuzumab has also been studied in SCC and AMl to target HGF (and c-Met).

In other cases of developments of resistance, ERBB3/4 and VEGFA, VEGFB overexpression is seen in these cases. Overexpression of PD-L1, PD-1, HAVCR2, TIM-3, and IDO also has been observed post anti-HER2 therapy.

Based on the studies (Analysis of Genomics biomarker actionability),[66] 96 percent of cancer patients have at least one target, identified in their genomic analysis either through tissue or liquid, consisting of a mutation 62 percent or amplification 29 percent (TP53 44 percent, KRAS 16 percent, or PI3K 12 percent). Eighty-five percent of patients have two or more alterations.

Amongst this 96 percent, *only 20 percent have an actionable target*, defined as an FDA approved/on-label drug. That said, using therapies directed through genetic analysis/biopsy has improved response rates for about 48 percent. This means that the more targets become actionable, we can expect a better outcome if therapies are customized based on this information.

As a result, there is a significant effort to enroll patients in the clinical trials directed through genetic information versus tumor type therapies. The largest precision medicine clinical trial is the NCI-MATCH trial. (For NGS trials, there is a cutoff of 16 percent for a promising therapy. The trial is patient sponsored!!)[67]

This trial looks at 11 mutations which could be treated through off-label use of targeted drugs. The concept also includes using these targeted therapies for secondary prevention of tumors sensitive to such therapies.

Another example is the Lung-MAP master screening protocol for NSCLC. In general, after the FDA expanded its access and right to try policies in 2020, it has been accessed by almost half of community oncologists, with about 70–80 percent success in obtaining access to unapproved therapies.[68]

The reality is that precision oncology is based on using off-label drugs, meaning that the clinical judgment of the oncologist plays the biggest role in choosing what drug, based on the genetic findings of a tumor, will be most effective. This is true even when using chemotherapies, and even when the genomic profile of the tumor is unknown. In other words, the concept of off-label drug use is conceptually old school, from before we knew anything about the genetics of tumors. One example is using Xeloda and Temodar for inpatients with high-grade neuroendocrine tumors.[69] Currently, using adjuvant therapies has no role in the treatment of neuroendocrine tumors and watchful waiting is advised in low-grade cases, or they can be treated with streptozocin.[70] What I have found is that topoisomerase II inhibitors generally are very effective in this setting, targeting the neuroendocrine-type tumors, but in the targeted world, surafatinib (which inhibits FGF and VEGF) has been suggested with some marginal efficacy (about 10 percent response rate). Other potential therapies include HIF-2 alpha inhibitors as well as peptide receptor radiotherapy.

Cyclin-dependent kinases, for example, were approved for breast cancer, and in different forms now are being studied in other types of cancer, due to the relationship they have with inactivating TP53. For example, in gliomas, NUV-422 is being studied as a CDK2/4/6 inhibitor. One study suggests that using CDK inhibitors is mainly effective when the AMBRA1 gene is not expressed.

Another example is erlotinib. It not only inhibits the EGFR, but at low dose it also inhibits EMT.

Erolotinib→HIF→hexokinase→EMT[71]

Erlotinib is also very effective in targeting the aerobic glycolysis, which is a very important metabolic target in CIMP types of cancer. CIMP plays a significant role in epigenetic control of hexokinase enzyme in colon CA as well as many others, such as glioblastoma.

In colon CA, CIMP is defined as hypermethylation of MINT1/2, MLH1, p14, p16 and lack of TP53. The presence of CIMP is parallel to MSI and BRAF mutations and the presence of right-side colon CA, mostly in women and older people, mucinous pathology, and poor differentiations.[72] These are all statistical differences, but not always significant. It is also reported to be associated with a hypermutated phenotype, but I disagree with this finding, as the epigenetically driven tumors are genetically stable. (There are two distinct types of CRC, chromosomal instability and epigenetic instability, or CIMP). In general, cases with MSI-H have a better prognosis. The pathology of origin is also different. CRCs originating from MSS (hypermutated) are originated from adenomas, and CRCs caused by CIMP-H are caused by epigenetic dysregulation originating from hyperplastic polyps.

The incidence of CIMP in colon CA is about 22 percent, and a good marker for stage II to consider avoiding chemotherapy, if positive. These tumors respond better to immune checkpoint inhibitors, as a general rule, due to the increased presence of lymphocytes inside the tumor. There is no association with history of smoking, diabetes, or NSAID use, but it is seen more often with alcohol consumption. This is of no surprise, as these risk factors (obesity, metabolic disease, smoking, etc.) are mainly mutation-driving risk factors and not methylation-driving risk factors, although there are studies on the effects of smoking on demethylating the histones (histone demethylating effect of smoking), but these effects could easily affect future generations.

I have a problem with the definition of alcohol consumption in this study, as it is defined as having 12 drinks in a lifetime, which I do not think should be considered as positive alcohol consumption. My standard definition of alcohol consumption is at minimum 10–20 grams per day. In other words, less than two drinks per day is mild, two to six drinks is moderate, and more than six drinks (60 grams) per day is defined as heavy drinking. In general, high alcohol consumption reduces the methylation of DNA and in my view is protective against many types of cancer. If the effects of DNA hypomethylation are significant enough to override the negative effects of general DNA hypomethylation, this is protective in the prevention of left-side colon CA. On the other hand, alcohol consumption can be associated with increased risk of right-side colon CA, as it is more dependent on global DNA hypomethylation. (Left-side colon CA is mainly CpG island hypermethylation and less dependent on global hypomethylation.)

BTK inhibitors are perfect drugs to be used off-label. Interestingly, ibrutinib is the drug that inhibits MEK and c-Myc. (SRC-1 expression in breast cancer is associated with HER2 and c-Myc expression and with reduced disease-free survival.) It is also shown that ibrutinib is able to inhibit EGFR twice as much as erlotinib![73] This effect is enhanced when ibrutinib is combined with auranofin. Auranofin is a gold compound typically used for parasitic infections and rheumatoid diseases. This medication is also effective in inhibition of macrophage tumor recruitment, and as such is a good candidate to be used for many other types of cancer, including sarcoma and neuroendocrine tumors.[74]

This old drug blocks Akt in synergism with ibrutinib. Based on its ability to inhibit thiols and thioredoxin and stimulate reactive oxygen species production, auranofin has shown promise as an anticancer agent in cancer cell lines including human leukemia, ovarian cancer, cervical cancer, lung cancer, and breast cancer. It has also been used in trials for CML and CLL.[75]

In one study it was shown to block the STAT3 in breast CA,[76] but in another study the gold did not impact the STAT3 in lung CA. Ibrutinib, however, blocks STAT3 in both cell lines. It also has been studied along with Torisel for use in lung cancer and ovarian cancer. It has been shown to have glycolytic activity similar to 3-BP, and is also suggested for GBM. In ovarian CA independent of TP53 it is effective (in case studies with some lab response). It is more effective in BRCA-deficient models, synergized with cytotoxic chemotherapy.[77]

Ibrutinib is also a key ER co-regulator in estrogen receptor-positive breast cancer. It is also suggested for use in advanced RCC, along with everolimus. I have treated very advanced cases of RCC with this protocol, along with epigenetic therapies, with results much better than the standard of care defined by Sutent. One major effect of this drug is to inhibit the macrophage recruitment of the tumor.

Another off-label drug suggested for prostate cancer is pegylated arginine deiminase, which is typically used for leukemias, but since prostate cancer has a deficiency in the urea cycle enzymes, this drug is effective.

In my practice, the targeted approach to prostate cancer consists of targeting the key differential genetic characteristics of the basal cells (TP53, RB1, c-Myc) and as such the therapies that most target these alterations will be considered in a metastatic setting, including epigenetic therapies as well as CDK inhibitors and mTOR inhibitors in combination as off-label therapy. Interestingly, studies have confirmed that basal cell genetic presentations, if confirmed as present in liquid biopsy, correlate with poor prognosis (JNCCN, Vol 19, 2021).[78] Just recently we have realized that skeletal metastatic disease progresses in the presence of ADT! Also, it is a major concern to me when practitioners use immune checkpoint inhibitors in a metastatic setting, again due to the negative impact reported in recent literature (JNCCN, Vol 19, 2021).[79]

In 1995 JAMA published studies confirming that the clinical judgment of the physician is superior to evidence-based medicine. This is the best example of this concept that a clinician can make—justification for using an off-label drug based on the genetics of the tumor—and this is exactly what is expected in precision oncology, where customized treatments are recommended.

Furthermore, trial designs should be, and I think will be, changed based on these basket trials for which next generation trials are designed. There are too many confounders playing a part, such as when the research studies are biased by their funding or when a simple hypothesis doesn't make sense. Studying, for example, if immunotherapy would be effective when combined with chemotherapy or radiation seems like an interesting concept. But you can look at the study looking at 3,112 patients with non-small-cell lung cancer, breast cancer, melanoma, colorectal cancer, or kidney cancer, who at the time of diagnosis had received definitive surgery for the primary cancer site but later developed brain cancer. This study showed that using immunotherapy with chemotherapy had no advantage over chemotherapy alone. By the same token, using immunotherapy with radiation and chemotherapy had no advantage over chemo and radiation alone.

Yet this study concluded that using immunotherapy with radiation alone had a survival advantage. This simply cannot be true, because when you are showing that immunotherapy does not add any advantage to chemo-radiation and it doesn't show any advantage when it's used with chemotherapy the simple mathematic logic (mathematical biology) tells you that immunotherapy with radiation cannot provide a survival advantage, unless you are to believe that chemotherapy and immunotherapy reduce survival if used in combination. Therefore, this study conflicts with itself.

Head and neck cancers in general are caused by HPV (where the p16 is mutated) or, if the HPV is negative, they are caused by smoking and alcohol consumption. The new approvals for immune therapies, including a new agent (toripalimab) to treat head and neck cancers, are again simply laughable. The data shows the response rate is 10 percent, and the older drugs that showed benefits, such as cetuximab, had better response rates (but still in the 10 percent range). So now they suggest combining them to treat head and neck cancers (with a 40 percent response rate). There has been suggested a similar combination of PD-1 inhibitors and targeted drugs in triple negative breast CA (famitinib and camrelizumab).

Further, other studies have suggested that immune therapies are more effective (if at all) when a patient with lung cancer is "obese." So I assume we have to overfeed patients with lung cancer so they can qualify for the study (Fox Chase Cancer Center study)!![80] That is suggested even though we know obesity is a high risk factor for cancers in general (breast CA especially). For example, increased levels of leptin and decreased levels of adiponectin are associated with carcinogenesis. This aside, much literature has shown increased risk of type 2 diabetes, hepatitis, and infections or colitis caused by immune therapies.

We also know that neurotensin is a marker we can detect in blood for cancer risk (with specificity of 90 percent). It is also a marker for obesity, and it is widely associated with increased cancer growth, as cancer cells carry neurotensin receptors at a high level. It is also used as a diagnostic tool in detecting imaging and is associated with reduced survival, as it increases IL-8.

Now, I understand that cachexia is a major problem in cancer, and lung cancer patients should be on optimal nutritional support and a healthy diet,[81] but the study at the Fox Center shows the response seen in patients who are actually obese, and average patients did not have the same response.

There are also studies confirming that a healthy diet improves survival in breast cancer patients (referring to the Dietary Inflammatory Index), or that high selenium protects against breast CA. The story of diet and a healthy lifestyle goes beyond the current common thought. As a matter of fact, as much as there is interest in promoting good health by changing one's life-style, I can testify that this is only the tip of the iceberg. For example, a new study suggests that a healthy life-style can mitigate high genetic cancer risk, using a cancer polygenic risk score formula (Elizabeth Rose in *Oncology Times*, Volume 43, Issue 20, Oct, 2021).[82] In this study the absolute cumulative risk of cancer was calculated based on environmental exposure and family history in a very precise way; however, the study is lacking in its fundamental point of not taking into account the epigenetic influences that are a dynamic phenomenon, besides the actual biases seen in the study subjects' reporting.

Let's go back to the misconceptions and misunderstanding related to the current standard of care. As we discussed about immune therapies, these treatments are approved when there is increased tumor mutation burden (TMB over 10, and in some studies over 16), but as we know, in high mutation burden tumors (such as colon and pancreatic cancers) the immune therapies do not work! This is the whole new concept of biomarker relevance. Even in melanomas, studies show that high mutational burden is only predictive of response when the affected individual is female! And no one knows why men with high mutational burden melanomas do not respond to immune therapies! It just makes me laugh! The prevalence of hypermutation also is dependent on geographic areas. For example, in China there is a very high incidence of lung CA, which is associated with MSI-High. The other known fact is that the presence of high mutational burden correlates with increased presence of actionable targets (PALB2, ERBB2, PI3KCA, TP53, ARID1, RB1, PTEN, NF1), meaning that the patient would benefit more from targeted drugs than immune therapies!

We also know that high mutational burden in breast cancer does *not* correlate with better response in breast cancer when immune checkpoint inhibitors are used (ASCO Vol 38, issue 22).[83] That said, pharma is still trying to find some market for their drugs in altered mismatch repair genes and BRAF V600E mutations in colon CA. In my opinion, this is a complete waste, as 20 percent improved PFS is a disaster for an expensive therapy with additional toxicity. Other studies suggest using adjuvant PD-1 inhibitors to improve recurrence-free survival in patients with RCC after nephrectomy, but if you look at the data it is pretty laughable, comparing 93.5 percent to 96 percent, with control! So out of 100 patients with M1 NED (no evidence of disease after nephrectomy and metastasectomy) or NED, only two of them will have a chance of better disease-free status after 24 months, but all 100 will experience toxicity!

In general, gliomas that have better biological features, such as IDH1, MGMT, TERT, or 1p/18q deletion, tend to have a better prognosis. That said, none of these groups respond to immune therapies. Also, although suggested, the neoadjuvant immune therapies for melanoma are not approved. Pretty much every time research studies talk about neoadjuvant immune therapies, I question, why try to use them before definitive therapy? Is it because they have a minimum role in an adjuvant setting and we simply want to hide the unknown factor?

That said, I was quite surprised to see a study that showed 82 percent response in patients receiving neoadjuvant immune therapies, in a wide range of patients, with a sample size of 24 (18 of whom had Lynch syndrome), mostly with CRC, after 6 weeks of therapy, measured by liquid biopsy (cDNA), which waived the requirement of surgery![84] If this study is valid, I would certainly question the statistical strength, and if it is not, I would keep it in my journal of comedies.

Worst of all are patients with so-called hyperprogressive disease where immune therapies paradoxically accelerate tumor growth and significantly shorten survival.[85] This is something I have seen many times in my practice.[86]

Also, as another example for using immune therapies in a subset of patients who are expected to do just fine, IMFINZI is promoted for unresectable stage III lung cancer after chemotherapy with cisplatinum, with 4 years median survival! No one asks, what about patients who have not received anything??

The immune therapies are also being suggested as neoadjuvant therapy for stages II to IIIB. Conceptually I still do not understand how immune enhancement is a type of complimentary therapy, where studies have even suggested that suppression of the immune system is possibly beneficial in treatment of patients with cancer (using indoximod with taxanes).

Another example of a nonrational use of immune therapies is the adjuvant use post chemotherapy in patients with stages IB to IIIA NSCLC for an "indefinite" course (at least 16 cycles).

Another example of failure for immune therapies is the combination of them with targeted therapies in melanomas, which has failed to show any superior results. That said, using imatinib in Kit+, melanomas has caused about 50 percent response.

Yet another laughable combination is bevacizumab, oral Cytoxan, and Keytruda all together in recurrent ovarian CA, suggested to improve PFS, where the combinations of two of these agents have not been tested. For example, oral Cytoxan has not been tested in this setting alone. This is when we already know that oral Cytoxan has a promising effect, especially combined with sodium bicarbonate in rat studies, for triple negative breast cancer, through its immunomodulatory effect without the need to use any immune therapy. As such, it seems that using immune therapies in this scenario, in ovarian cancer, simply is unnecessary and would have no role in the outcome. Since 80 percent of ovarian cancers are metastatic at the time of diagnosis, the immune system's access to the tumor antigens is a given. Another example of immune therapy in ovarian cancer is maveropepimut to enhance T cell activity, currently being studied in breast and ovarian cancer.

Another example is that in brain metastatic disease the addition of PD-1 inhibitors did not improve survival when it was added to radiotherapy, but the study was called positive, because of the positive response rate!

I wanted to make this point because when we are looking at this sort of trial, with a large pool of patients with different types of cancer that simply uses a very wide range of genomics mixed in the pool of the systemic review or the prospective trial, the data becomes so confounded that it cannot be used for any single patient who is hoping to use the recommendations coming from the trial. The second problem with the majority of the trials is that the primary endpoint is not survival; rather, it is the objective response according to the RECISTversion 1.1 criteria.

When a primary end point of the trial is not survival, it tells you that the hypothesis was lacking enough evidence to go after survival; then as such the primary endpoint used is either a response rate or progression-free survival, which can at many points fail to translate to better survival. In fact, progression-free survival can be a marker for response only for daughter cells, versus overall survival, which is a marker for response from stem cells.

Then finally, many of these trials have looked as the RECIST criteria as the primary endpoint, where the radiological findings can be completely irrelevant to the patient's survival. Objective responses, for example, on patients who had received targeted therapy at the very end stage, showed only reduced tumor burden, and no advantage for patients' outcome. This could be true also for some targeted radiation therapies, such as Rhenium Nanoliposomes targeting glioma cells and delivering targeted radiation.

Studies also have confirmed that using proton therapy is superior to electron therapy (external beam or brachytherapy) in its efficacy in prostate cancer, on both overall survival and tolerability (Clinical Genito Urinary cancer, Vol 19, Issue 3, June 2021).[87] The practical application of proton therapy is indeed underappreciated, mostly due to reimbursement issues. In my clinic, patients who receive proton therapy are referred by a proton center in San Diego for epigenetic therapies simultaneously, and the reason is clear. The intratumoral hypoxia is reduced with the application of histone deacetylase inhibitors, as a result of inhibition of hexokinase. This phenomenon has resulted in shorter duration of proton therapy required to obtain similar results, as well as reduced radioresistance and improved response overall. Currently, patients enjoy reduced collateral damage (especially when the radiation is applied to vital organs, such as the brain to treat brain tumors, with reduced incidence of cognitive impairment, compared to standard therapies), as well as improved outcome.

Another example of these response rate trials with so-called promising results used a PD-L1 inhibitor such as Keytruda for treatment of refractory solid tumors with PD-L1 as status of positivity, which in fact was one of the main criteria that made practitioners believe that the solid tumor should respond to this immunotherapy. The therapy performed on 1,050 patients only showed a 9 percent response based on the RECIST criteria, again in patients with PD-L1 tumors. In other words, 91 percent of these refractory solid tumors did not respond to Keytruda alone, even with positive PD-L1, and as such the investigators concluded that high tumor mutational burden status has to be considered as a positive factor if the patient is treated with these agents. It is notable that in in PD-L1 negative tumors, the response rate was only 3 percent, yet we see day after day that patients with refractory solid tumors with unknown PD-L1 status or negative PD-L1 status are treated with this agent. As I said, in PD-L1+ patients the response rate is only 9 percent, let alone those who are negative, who have a less than 3 percent response. The general mechanism of resistance to PD1 inhibitors includes FGFR3, TP53, and RB1.

RB1 is also a very important mechanism of resistance to cyclin-dependent kinases. As such, selective epigenetic therapies (demethylators) are very effective if combined with cyclin-dependent kinases to inhibit the mechanism of resistance. In my practice breast cancer patients are treated with a combination of epigenetic therapies along with cyclin-dependent kinases to avoid this phenomenon (as well as the inhibition of FGFR, which overexpresses with hormonal inhibition). Using everolimus in sequence with CDK inhibitors is also important, as the mTOR is upstream of CDK. One of the main reasons a patient develops resistance to CDK inhibitors is inactivating mutations of RB1, and this is in concert with our rationale on using demethylating epigenetic therapies to enhance the RB1 expression. One of the other very common sequencing strategies is the inhibition of the EGFR and mTOR pathways, as one reverses the other mechanism of resistance.[88]

The effects of the Covid-19 pandemic have caused many significant shifts for the worse in the care of patients. These generally include increased hospitalizations and reduced capabilities of medical care. For example, the death risk for cancer patients is reported at 5–50 percent. The risk of death is four times higher in patients over the age of 70 with lymphoma when affected by Covid. That said, immunization is ineffective 85 percent of time in this group of patients (with 16 percent effective immunization in patients with CLL and multiple myeloma on active treatment)[89] and 55 percent in treatment-naïve patients.

Talking about the side effects of Covid-19 vaccinations is beyond the scope of this review. I just would refer to several of my own publications in regards to adenovirus vector Covid-19 vaccines and concerns in relation to thrombotic thrombocytopenia,[90] carcinogenesis, and immune disruption. Keep in mind that cancer patients always have less antibody production, and in patients who have received chemotherapy in the last 12 months the antibody production is 24 percent, compared to 69 percent in chemo naïve patients (JAMA Oncology Nov, 2021).[91]

There at least seems to be one positive factor that has resulted in perhaps a better understanding or accommodation of patients who are dealing with cancer and are not particularly benefiting from the therapies. In other words, the benefit to risk ratio of the cancer therapy has shifted at this time, and we are now recognizing that for patients with metastatic disease treatments only have small incremental benefits to survival, and during this time of pandemic these treatments should be reconsidered and a supportive care approach should be prioritized. This is a shift away from the current or recent therapies that have been advocated with only very small advantages for patients with metastatic disease; in other words, through lack of resources it seems that we are using them more wisely. Also for newly diagnosed patients we have seen a trend in postponing their surgery (in the case of women with breast cancer) and providing an option for neoadjuvant hormonal therapy for four months (the IMPACT trial), in an attempt to convert the noncandidates for breast conserving surgery into candidates. It is proposed that such a strategy could also reduce the chance of recurrence of disease by the reduction of Ki-67 (and predict this possibility if the Ki-67 stays higher than 10 after 2 weeks of aromatase inhibition). Obviously my argument would be that patients would not choose to be biopsied twice in 2 weeks to measure the Ki-67, and furthermore it is not expected that Ki-67 will drop to less than 10 in 2 weeks if the initial finding is higher than 20, which most of the cases are. This type of so-called molecular de-staging is also something I have been doing for over 10 years, using epigenetic therapies instead of hormonal blockers. Measurement of Ki-67 has proven to reduce post epigenetic therapies. One other strategy recently suggested includes using cyclin-dependent kinases, along with AI, with companion prognostic tests such as the 4 gene assay EndoPredict (by Myriad).

One other effect from the Covid infection is that patients with cancer continue to shed viruses over 78 days post diagnosis. That said, one wonders how the recommendations for treating these patients have not changed. (In our clinic, patients with Covid-19, with or without cancer, are expected to stop shedding in 3–4 days.) But overall the Covid pandemic has caused less administration of chemotherapy and as a result 70 percent of postmenopausal patients who receive neoadjuvant hormonal blockade instead of chemotherapy have shrinkage of their tumor (these are patients aged over 50 with an Oncotype DX score of 0–25). The lack of benefit was also seen in patients over the age of 50 who had positive LNs, up to three, with scores less than 25.

The Oncotype DX score only pertained to ductal carcinomas, not lobular. Other molecular profiles include MammaPrint, Blueprint (which accurately predicts the CR post therapy in all subtypes of breast CA—luminal, basal, HER2…), EndoPredict, and the Breast Cancer Index.

In a discussion about precision oncology it is important to understand what the term actually means. Precision oncology means "one suit does not fit all." In other words, it disputes the rationale of conducting clinical trials that are based on tumor type. It suggests that the treatment of any cancer patient should be different than the treatment of another. If this is the case, the use of traditional therapies which are approved based on systemic reviews and the highest evidence-based medicine (meta-analysis) would become irrelevant. That said, tumor type still plays a role. Keep in mind that the response rate in patients with a BRCA mutation is tumor type specific, as only 14 percent of patients with lung cancer, for example, respond to a PARP inhibitor when this is mutated.

The concepts of basket trials, next generation trials, and biomarker-based therapies are all in conflict with standard of care practice. Understanding the cellular biology has provided us with tools to be able to differentiate the responders from the nonresponders. For example, again in pancreatic cancer, the presence of a glycan protein, sTRA, not only can allow us to detect the cancer long before the CA 19.9 detection, but also can differentiate the patients who will respond to chemotherapy from the ones who will not. This protein is expressed in 65 percent of patients, meaning that two thirds of patients will not be candidates for chemotherapy. If they do respond to the therapy, options including Xeloda, olaparib, and FOLFOX can be used as maintenance therapies. There are three approved PARPi for maintenance: olaparib, rucaparib, and niraparib. All three have an enhanced effect when combined with a DNMT inhibitor.[92] This preclinical data further was clinically supported by my clinical data and related approved patents. Niriparib has also been studied in gliomas.

In pancreatic cancer, using FOLFOX after Gemzar is very good idea. Isn't it? Studies show the response rate is 4 percent and the duration of response (PFS) is 2 months. In general patients have 4–5 months as OS, with a very poor quality of life. The other therapies looked at for pancreatic cancer obviously include vitamin D, IL-6 inhibitors, and inhibitors of CXCLI2. A recent study also showed a better response rate with liposomal irinotecan in FOLFIRI, versus traditional FOLFIRI (19 percent versus 9 percent). Other studies using targeted drugs in combination with chemotherapy have also failed (for example, dasatinib plus chemo). This is the case specifically with squamous cell nature (basal-like) tumors that express KRT5, KRT17, and TP63. The luminal types express GATA6, TFF1, and CLDN18. Onvansertib (a PLK-1 inhibitor) recently has been suggested, along with irinotecan.

In my opinion, the real burden of the disease is mainly compared to the burden of treatment toxicity, and treatment success can not be assessed based only on tumor response. That said, it amazes me how case studies such as the one published in JCO in Oct 2021, by Singh et al,[93] is presenting successful treatments for patient with pancreatic cancer who have received 74 cycles of chemotherapy!

One major discussion in drug development is the term *quality adjusted life years*. QALYs were developed as an outcome measure to incorporate the effects on both the quality (morbidity) and quantity (mortality) of life. Each year of life is multiplied by a weighting factor reflecting quality of life. An alternative to QALYs is the healthy years equivalent, which is discussed below.

QALYs are estimated by assigning every life year a weight between 0 and 1, where a weight of 0 reflects a health status that is valued as equal to being dead and a weight of 1 represents full health. For example, consider a patient with colon cancer who has a health state of 0.9. Without surgery, he will die in 2 years. With surgery, his health state deteriorates slightly to 0.7, but he lives 5 more years (in total 7 years). Therefore, the QALY gained with surgery = $[(0.7\times7) - (0.9 \times 2)] = 4.9 - 1.8 = 3.1$.

There is, however, significant effort to minimize the value of the clinical utility of biomarker-guided therapies. Tumor biomarker tests are used to direct care of patients with or at risk for malignancies. In the era of precision medicine, TBTs have become increasingly important given that their results may be critical determinants in the decision to proceed with, or withhold, effective therapies that may have substantial toxicities. Taken together, these circumstances have resulted in concern about the true clinical and financial value of a TBT and, therefore, its clinical utility.

In 2009, the Evaluation of Genomic Applications in Practice and Prevention Initiative proposed three semantics to determine if a genetic test should be used to manage care: analytical validity, clinical validity, and clinical utility, which have been expanded to refer to any diagnostic test. This section of the book will relate only to TBTs, and these terms have been adopted by both the NAM and ASCO for deliberations specifically regarding TBTs. Controversial definitions of these three terms have led to confusion among investigators, regulatory and guideline bodies, clinicians, third-party payers, and other stakeholders about when a TBT should, indeed, be introduced into clinical care.[94]

This simply is a result of mismanaged information from biomarkers, when it comes to prioritizing them in an order that correlates with patients' survival. The main problem with designing any trial or for that matter therapy around the biomarkers, is that the biomarkers selected need to be functional and not molecular alone.

For example, in the Beat AML Master clinical trial, patients are treated based on the biomarkers identified instead of the chemotherapies used such as cytarabine and daunorubicin, which have been used for the last 40 years. One finding is that when a patient with AML carries the IDH1 mutation, the addition of venetoclax to azacytidine improves survival. Ivosidenib and enasidenib also have both targeted this mutation effectively. The other combination is navitoclax and ruxolitinib. Also we know that in AML, assessment of the leukemic stem cells biomarkers CD34+, 38- and TIM3+ is suggested to predict the response after induction therapies.

As we discussed earlier, the IDH1/2 activity in gliomas are parallel to the Warburg effect and CIMP phenotype. As such, using epigenetic therapies is suggested.

Iadedemstat blocks the LSD1, which is a main target for hematological cancers (AML) as well as Ewing sarcoma. LSD1 is a known target for its influence on the glycolytic system in leukemias (LSD-1 activates this pathway by inhibiting GATA1). Pevonedistat is an inhibitor of NEDD8 and approved for AML and high-risk MDS, and Curis's CA-4948 (an IRAK-4 inhibitor) is also approved for AML. Sabatolimab is an anti-TIM-3 monoclonal antibody suggested for MDS/ AML and CML.

Most of the currently approved drugs are based on a combination of a type and target specific indication. Also, gene therapy has become more realistic (in the cases of sickle cell anemia and thalassemia).

CD34→VECTOR→m RNA

Examples here are ex vivo autologous hematopoietic stem cell gene therapy with a retroviral vector, and lentiviral self inactivating therapy.

For example, sotorasib is indicated for NSCLC with KRAS G12C mutation (along with many other TKIs, such as CDK, mTOR, PD-1, MEK, and even chemotherapies). KRAS mutations are seen in 20 percent of lung CA, 100 percent of pancreatic CA, and, depending on oxygen saturation in colon cancer, up to 100 percent. KRAS mutations are transition or transversion. Onvarsertib and selumetinib are also KRAS inhibitors. The issue again is that CTC patients with mutated and nonmutated KRAS have the same median survival.[95] One of the mechanisms of resistance to MEK inhibitors is through FAK/ PYK2, and as such the use of Defactinib (an FAK inhibitor) in this setting is being studied in combination with MEK inhibitors, starting with low-grade serous ovarian cancer, which is 70 percent KRAS+. Traditionally these tumors were treated only with hormonal (ER) inhibitors, such as letrozole. Other options include folic acid targeting antibody-drug conjugate.

APX005M, however, as a CD40 inhibitor, affects tumor growth regardless of KRAS state when used in pancreatic cancer.

The most common mutations in CRC are APC (72 percent), TP53 (47 percent), KRAS (40 percent), and PI3KCA (16 percent). There are studies suggesting categorizing colon CA based on a consensus molecular subtype

(CMS), with classifications of C1: Hypermutated, C2: Wnt activation (which recruits macrophages and causes malignant ascites), C3: Metabolic, and C4: Mesenchymal. These classifications are simply meaningless, as in different stages of tumor growth the tumor develops and chooses to use different classes of drivers and behaves differently. It is only a made-up theory that if we find one class that could potentially respond to a drug, we can use the related biomarkers as companion diagnostics and get approval for the drug! For example, the growth factors CEACAM1, ANG-2, and ICAM-1 have been suggested for the same reason to get approvals for the drug nintedanib. Interestingly, none of these markers showed any correlation with the response to this drug.

The application of molecular subtypes or genoproteomic characterization is one way that researchers again try to put a number of patients with each type of cancer into a group and show that certain medications or approaches are more effective in that certain defined group. So far eight types of cancer have been formulated in this classification. One example, as we discussed, is CRC. Others include TNBC and HPV head and neck cancers. In this type of cancer, CIN refers to a subtype where there is overexpression of cyclin-dependent kinases. The basal subtype, however, is the type that overexpresses the growth factors, such as EGFR and TNF-1. Finally, the immune class is the one that is treated with immune therapies.

Studies on targeting the RAF/MEK pathway are very conflicting. For example, the presence of an MEK/MAPK mutation is a mutually exclusive event from other mutations (except TP53/APC) and excludes the benefit of using an EGFR inhibitor. The ERK inhibitor urixeltinib also has been deemed ineffective in this setting. That said, only a few cases have been studied to present this theory and yet BRAF inhibitors are used in this setting with no reasoning. Also, the suggestions that the BRAF mutation is more common in the setting of MSI is not realistic.

Tranexamic acid is a anti-bleeding medication usually reserved for patients with myelofibrosis or acute bleeding after trauma, and it has shown strong anti-EGFR efficacy (1300 mg orally for 12 days).

One would ask, if the drug is effective on the target, why should it be limited too a specific type of disease? These two concepts are treated as mutually exclusive, as if a drug is approved and effective for one target, the type-specific application is disputed by the concept, and if it is approved and effective for a tumor type, the target-specific indication is irrelevant.

One very useful test looking at fusion mRNA and DNA mutations is from the Genomics Lab Cooperative [96] but the challenge remains, as many times, the gene overexpressions and activating mutations do *not* translate to increased protein expression or vice versa, as such identification of dysregulated oncoproteins is important. "A fraction of tumors are found with aberrant activity of druggable oncoproteins despite a lack of mutations, and vice versa. In vitro assays confirmed that VIPER-inferred protein activity outperformed mutational analysis in predicting sensitivity to targeted inhibitors."[97]

Also, Dr. Andrea Califano's team at the Columbia University has developed an interesting new tool called Multi-Omics Master-Regulator Analysis to analyze the genetic alterations in the tumor and identify the master regulators. This process has identified the growth factors to be key regulators in carcinogenesis. For example, in Ewing sarcoma, the drug TK-216 was developed to inhibit the IGF-1R with a very sustainable outcome, and MPO250 to inhibit VEGF and HGF in solid tumors. Another new drug for sarcoma is Annamycin, which is a next generation doxorubicin. In rhabdomyosarcomas a combination of irinotecan and Temodar seems to be superior to irinotecan alone. Most patients refrain from vincristine due to its neurotoxicity. In my view chemotherapies have a minimal role in treating sarcomas in general, especially certain types of sarcomas, such as leiomyosarcomas. Recently surafatinib was developed to target VEGF and FGF (in NE tumors).

We have focused most of our attention on the activity of oncopromoter genes and yet we know that their activity is in concert with silencing or loss of function of oncosuppressors, for example AMBRA1 loss of function correlates with CDK activity or loss of function of CUX1 leads to development of AML.

Unfortunately, except in some rare instances we do not have active drugs for loss of function, and therefore our hands are tied in re-establishing the function of oncosuppressor genes. In general this concept (loss of heterozygous state) is far less significant when you take into effect the impact of epigenetics on gene function, as somatic events can serve as a second "hit" to these genetic signatures. As such, the epigenetic silencing of the gene can cause biallelic inactivation.

That said, using natural compounds has given us the ability to effectively turn on the genes that are silenced and overcome the inactivating mutations in oncosuppressors. A very good example of that is using PTEN activators in patients who have a mutated BRCA gene. Although in the therapeutic field this could be tricky, as you can reestablish the DNA repair genes at all times, this is in contrast to using PARPi when a patient with BRCA genetic mutation is treated. This causes a double-edged sword effect in the therapy, as now the cancer cells can survive with the re-establishment of the oncosuppressor activity. I personally use specific protocols in these circumstances, where the gain of function for one oncosuppressor is needed to maintain disease free-survival; at the same time, the tumor cells are targeted with antineoplastic agents, specifically targeted therapies. A nice admission by researchers at Florida A&M University (by Samia S. Messeha) showed an effective compound, named sanguinarine, as able to induce apoptosis in breast CA, along with other anticancer activities, in many other types of cancer, leukemias, and so on.[98]

Targeted therapy directed by genetic testing in patients with advanced refractory solid tumors, lymphomas, or multiple myeloma, *match screening*, where the patient receives targeted therapies based on the alteration found, is based on the same concept but is in the very early stages.[99]

There are three categories of drugs used in MM: proteasome inhibitors, anti-CD-38 antibodies, and immunomodulatory agents. Blencyto can also be used. Proteasome inhibitors are now being studied in all other types of cancer; for example, ixazomib is being studied along with Faslodex for use in breast cancer. (This is a drug used in multiple myeloma in a maintenance setting.)

To give another example, patients with positive CD30 (Hodgkin's lymphoma) are treated with brentuximab (which is a monomethyl auristatin conjugate to CD30 targets), or patients with diffuse large B-cell lymphoma are treated with tafasitamab (Monjovi) with lenalidomide (for CD-19+), or glofitamab (for CD20 and 3). All represent bispecific inhibitors.

Another option for these patients is a recently approved drug, polatuzumab, along with bendamustine and rituximab (which nowadays comes in SC form). Generally, the activated B-cell subtype is resistant to chemo-immune therapies, relative to germinal center BC, due to activated Bcl. Other options for DLBCL include selinexor and mosunetuzumab. The reasoning behind the resistance of lymphomas to immune therapies in general is that it could be due to the activation of JAK1/2 and reduced beta 2 microglobulin. The presence of passenger mutations at subclonal levels is consistent with such resistance to the therapy. This is in contrast to the fact that development of passenger mutations in a normal individual does not correlate with malignancy, as 99 percent of mutations are passenger.

Patients with refractory gliomas also can be treated with MDNA55 if their IL-4 is overexpressed, with an additional 10 months survival in a refractory setting. Another glioma treatment is HSV-1 G207 immunovirotherapy for pediatric gliomas. Placenta-derived NK therapies are also now approved for ODD status in gliomas, and in pediatric gliomas, mutations in ALK, NTRK, ROS1, and MET have been explored.

One other definition of precision oncology is to use off-label products for each cancer type. The use of drugs for indications beyond those stipulated on the FDA label has been a physician prerogative that the FDA has approved of for more than 30 years. A high percentage of the use of drugs and biologics in cancer care is off-label. The introduction and utilization of many innovative drugs and biologics in oncology practice, combined with their significant expense, has focused the attention of payors and other involved constituencies on processes and programs that facilitate the appropriate, effective, and efficient use of such agents. Since the late 1980s, drug compendia have been one mechanism to arbitrate such appropriate use in terms of both payors and providers. One area of interest is de-escalating the treatments, along with the need to design de-escalating trials.

On December 17, 2010, NCCN convened the NCCN Oncology Policy Summit: Off-Label Drug Use and Compendia in Washington, DC. The NCCN Oncology Policy Summit examined the use of compendia by public and private payors, looking at issues such as categories of evidence, compendia processes, and availability of data, among others.

There is significant effort put into making new drugs, with the idea that new delivery systems make them work better. One example is drug antibody conjugates (DM1-ICAM1) for pancreatic cancer, or nano RNA with paclitaxel or irinotecan, or a newly approved drug, Trodelvy, for triple negative breast CA, and ENHERTU (fam-trastuzumab deruxtecan, or T-DXd) for HER2+ breast CA (although with a minimum PFS advantage of 6.9 months and response rate of only 45 percent). Guardant 360 is being considered as a companion diagnostic for this.

ENHERTU and zanidatamab are also approved for gastric CA with HER2+. The concern is about this drug's interstitial lung disease (tucatinib does not have this side effect). Another drug antibody conjugate recently studied is called SORTOLIN ab- docetaxel. Another one is enfortumab vedotin (targeting nectin-4), oportuzumab, ASG-15ME, DS-8201 alpha, disitamab, T-DM1 (in urothelial cancer), and other types. Datopotamab deruxtecan is also a TROP2 and TOP1 combo, used in lung cancer. Tusamitamab is an antibody drug conjugate under development for use in lung cancer.

In stage IV triple negative breast CA, only PD-L1+ patients benefit from immune checkpoint inhibitors. There are ongoing studies on using RU486 plus enzalutamide. Estrogen has been used on patients with triple negative breast cancer exhibiting ERbeta. Also ERbeta agonists can be used to counter the ER resistance mechanisms by upregulation of FOXO transcription. This is the mainstay of using *natural* estrogen-like compounds, such as sage and licorice, as they counter the resistance produced by using synthetic estrogen blockers, such as tamoxifen and AI. In my practice, patients who develop resistance to estrogen inhibition and develop an ESR1 mutation, are *not* treated with Faslodex, as it is commonly used in ligand-binding domain mutations; many of these resistance mechanisms are induced by chimeric fusion proteins that contain the N-terminal sequences of ER alpha encoding the DNA binding domain, fused to the C-terminal sequence to form a partner gene that activates transcription without binding to estrogen.

Talazoparib also has shown a better outcome compared to chemotherapy (with a median OS of 8 months compared to 5 months). The addition of PI3K inhibitors has been shown to improve response to the PARP inhibition, as well as the androgen inhibition. Ipatasertib and capatasertib are not proven to improve OS. The ideal approach would be to look at the subtype of the disease (there are six subtypes in TNBC: two basal-like, immune modulatory, mesenchymal/stemlike, mesenchymal, and finally lumina). Another more recent drug for TNBC is ladiratuzumab (LIV-1 antibody).

One other example is folic acid receptor targets with PARPi (in ovarian CA and TNBC): mirvetuximab and rucaparib, or Avastin (obviously the concept is deficient, as folic acid receptor binding will activate the aberrant epigenetic markers). We ourselves developed folic acid targeted therapies aimed at ovarian cancer expressing these receptors and were surprised to see that ovarian cancer grows faster when these targeted drugs are used. It was our experience that led us to believe that the binder itself targeting the folic acid receptor activates an undesirable epigenetic influence by the folic acid effect. As such, I am quite pessimistic about the combinations of mirvetuximab in the standard research setting to treat ovarian cancer.

Obviously sacituzumab govitecan (Troveldy) has been approved and is effective in the setting for HR+/HER2- tumors as well, as they all carry TROP2 (50 percent expression), but is only used for four cycles, with an expected PFS of 5.5 months. Trodelvy is also approved for urothelial cancer.

Here I would like to review a series of off-label products and their mechanisms of action:

Off-label drugs used in cancer:

1. metformin and doxycycline: We do not use it. It is an mTOR inhibitor and inhibitor of mitochondrial phosphorylation/respiration. That said, I am interested in the drug devimistat (developed by Rafael Pharmaceuticals), as it appears to be a good candidate for mitochondrial switch when tried in pancreatic cancer. Pevonedistat is also approved for AML (E3 ubiquitin ligase). (Devimistat inhibits both pyruvate dehydrogenase and alpha-keto glutarate dehydrogenase.)

2. tetracycline + vitamin A/ C: strong demethylator of TET.

   It is important to know that under hypoxic conditions, TET/ HIF form a combo molecule to increase tumor aggressiveness and proliferation. *As such the activation of TET only should be suggested after the HIF is inhibited.* Hypoxia, Epithelial-Mesenchymal Transition, and TET-Mediated Epigenetic Changes.

   Hypoxia→HIF→TET→HIF

   MK 6482/belzutifan inhibits HIF-2, and received breakthrough designation by Merck, for nonmetastatic RCC (less than 3 cm), alone or in combination with lenvatinib, ± PD-1 inhibitors.

HIF is degraded by VHL. Ninety percent of RCC cases have defects in VHL. It is also common in pancreatic neuroendocrine tumors (approved indication for belzutifan).

3. perphenazine: It inhibits the mechanism of resistance to BRAF and EGFR

   https://www.ncbi.nlm.nih.gov/pmc/articles/PMC6720634/ (BRAF inhibitors only have a 20 percent response rate on BRAF + colon CA).[100]

   This is important, as 50 percent of EGFR tumors develop T790M mutations post targeted therapies.

4. DOCK1 inhibitors.

5. FATP2 inhibitors: Inhibitors of metabolism of fatty acids, they prolong the response to BRAF inhibitors.

6. SFRP2 inhibitors: SFRP2 is involved in the Wnt pathway and angiogenesis.[101]

7. mTOR resistance can be overcome by inhibition of CDK9, and c-Myc (mTOR) inhibition activates the cancer stem cells by setting off the alarm among homopoietic stem cells. In response, they proliferate, which produces rewired cancer stem cells using new signaling pathways that render mTOR signaling pathways.

8. valproic acid 250 mg once a day

9. promethazine 25 mg once a day

10. hydralazine 25 mg once a day

11. entacapone 200 mg a day

12. digoxin 125 mcg a day

13. amiloride 5 mg bid

14. rifampin 300 mg once a day

15. nelfinavir 1250 mg twice a day X 7 days

16. disulfiram 500 mg once a day

17. Bactrim once a day

18. riluzole 150 mg bid

19. cyproheptadine 4 mg per day

90. Propranolol 20 mg per day

21. Plaquenil 200 mg a day. The use of *antimalarials* in cancer has been studied widely due to their effect on inhibition of mitochondria as well as inhibition of tumor hypoxia. (The tumor's oxygen consumption would be reduced.) Plaquenil specifically targets PPT1 and controls mTOR.

22 Valcyte (Valgancyclovir), 450 mg bid

23 artesunate 50 mg po twice daily

24 aprepitant 80 mg po twice daily

25 sertraline 50 mg po twice daily

26 captopril 50 mg po twice daily

27 auranofin 3 mg po twice daily

28 nelfinavir 1250 mg po twice daily

29 temozolomide 25 mg/M2 po twice daily

30 disulfiram 250 mg po twice daily

31 copper (cupric) gluconate 2 mg po twice daily

32. ketoconazole 200 mg po twice daily

33. Antirheumatoid drugs: Gold (auranofin) 3 mg twice a day up to 21 mg per day (for 14 days) inhibits telomerase activity. The side effects are not significant.[102]

The use of antiparasitics in triple negative breast CA has also been studied (by Peter Lee, MD), alone or in combination with PD-L1 inhibitors, with some results in the animals (ivermectin modulates the P2X4).

Teratogens: To the common scientist's mind, this at times could be meaningless, as the common person with common knowledge may not see the connections existing between different findings beyond medicine. One good example is the use of teratogens in oncology, a concept that I came up with after understanding the stem cell mechanisms of repair and reproduction. One of these pathways, known as the hedgehog pathway, has long been studied and a recent medication called glasdegib approved for and used in AML, basal cell carcinomas, and so on. This target is activated after an animal is poisoned by grazing toxic plants or by eating feed that is contaminated by toxic plants. Affected livestock often have compromised reproductive function including infertility, spontaneous abortion, and fetal deformities. The same concept applies with human cancer stem cells and the application of "homeopathic doses" of the same compounds could be a strategy to fight human cancer stem cells. In detail, I use this approach in my clinic (e.g., by using Vertara in leukemia).

The application of teratogens in patients with cancer is based on the concept of stem cell treatment. If we believe that stem cells are the cause of cancer, use of teratogens can reduce the synthesis of the tumor. Below is a list of known teratogens, many of which have been used in my clinic:

- angiotensin converting enzyme (ACE) inhibitors, such as Zestril and Prinivil
- aminopterin
- carbamazepine (Tegretol)
- chlorobiphenyls
- danazol (Danocrine)
- etretinate (Tegison)
- isotretinoin (Accutane)
- lithium (Eskalith), specifically used to inhibit glycogen sythetase kinase and E-cadherin induction through Wnt induction
- methimazole (Tapazole)
- methotrexate (Rheumatrex)
- penicillamine (Depen, Cuprimine), specifically used in melanomas
- phenytoin (Dilantin), specifically used in melanomas
- propylthiouracil (PTU)
- prostaglandins
- tetracycline (Sumycin)

- trimethadione (Tridione)
- valproic acid (Depakene)
- coffee (reduces the risk of melanoma and colon CA)

Lymphoma maintenance meds

- hydralazine 25 mg twice a day, a specific demethylator agent
- indomethacin 25 mg twice a day
- valproate 250 mg a day
- promethazine 25 mg twice a day
- Bactrim, one single strength tablet, once or twice a day, a demethylating drug

Breast CA maintenance meds

- Bactrim 1 tablet a day
- hydralazine 25 mg twice a day

Pancreatic CA maintenance meds

Tamoxifen, Plaquenil, metformin, doxycycline, olmesartan, atorvastatin, meloxicam, loratidine, mebendazole. Doxycyclin is suggested to differentiate the stem cells in studying the new concept of turning on the PTF-1 gene.

Chapter Eight

Mutation Analysis and Epimutations

I WANT TO ADDRESS THIS IMPORTANT SUBJECT OF GENETIC AND epigenetic heterogeneity, and the association with germline mutations predicting somatic mutations. It is important to understand if genetic aberrations have occurred at different stages of clonal differentiation. Some genetic mutations happen in very primary clonal differentiations and some are late in subclonal differentiation. Many of the cancer somatic mutations are associated with other diseases as well. For example, mTOR mutation has been reported in childhood seizure disorders, or KRAS mutation has been reported in AV malformations. Also, macrophages carrying TET mutations are proinflammatory and can accelerate plaque destabilization. Patients with autoimmune disorders and MS can have STAT3 mutations, explaining the high rate of these types of diseases in cancer patients.

As I will explain later in this book, the epigenetic role in carcinogenesis at a younger age is well documented in genetic studies of oncosuppressor genes. It is estimated that by 2030 the incidence of colorectal cancer will increase by 124 percent in patients aged 20–34. This is besides the known FAP, or Lynch syndrome incidence, as we know these syndromes come with increased risk of other types of cancer, such as gastric, thyroid, and liver cancers.

One evident mechanism in early-onset colon cancer is due to a constitutional MLH1 epimutation transmitted through the maternal germline from a mother with mosaic, dense hemiallelic MLH1 methylation in only a small proportion of her cells. The epimutation can occur in a promoter without apparent genetic alterations. Such constitutional epimutations can be screened and if positive (such as constitutional hypermethylation of the MLH1 promoter), it appears reasonable to evaluate their parents for evidence of gonosomal mosaicism. This low-level constitutional hypermethylation also appears in other oncosuppressor genes, such as BRCA1 or CDNK2A. The illustration on page 109 shows the relation between the non-Mendelian inheritance of the affected individual.

In the below reported case of constitutional MLH1 epimutation, no recombinant event accounting for the non-Mendelian inheritance of the epimutation, including SNPs, could be found.

The findings in this case can be explained by a de novo event occurring in an embryonic stem cell in the mother as she developed in utero. That abnormal embryonic stem cell gave rise to the dense methylation and loss of expression seen in approximately 5 percent of her soma and germ cells, whereas her remaining cells derived from unaffected embryonic stem cells showed normal MLH1 expression. This "gonosomal mosaic" model predicts that the proband arose from fertilization of a rare oocyte with the epimutation whereas another sibling can inherit the same haplotype from an oocyte without the epimutation, resulting in non-Mendelian inheritance. Nevertheless, this phenomenon has clear implications for assessing cancer risk in families in which individuals show low-level promoter hypermethylation of cancer predisposition genes in constitutional epigenome.

The detection of germline mutation can be in either blood or tissue. The correlation of somatic mutations to germline mutations is very interesting, as it can guide the preventive measures in the family. For example, presence of somatic mutations in a patient with a BRCA+ tumor can or cannot accompany germline mutations. Regardless, if the application of PARP inhibitors is also indicated, the response is always stronger with germline mutations. Currently PARP inhibitors are even used without the presence of such mutations in certain circumstances. (The concern would remain at the end about the risk of development of MDS in patients receiving this drug for a long time. Obviously, for the pediatric population with MDS the story is different, and mainly has to do with GATA and SMAD mutations.)

It is also important to know that the metastatic tendency of a tumor is predefined by mutations happening in the early stages of development, except in the case of RCC, when it happens late in development. One of

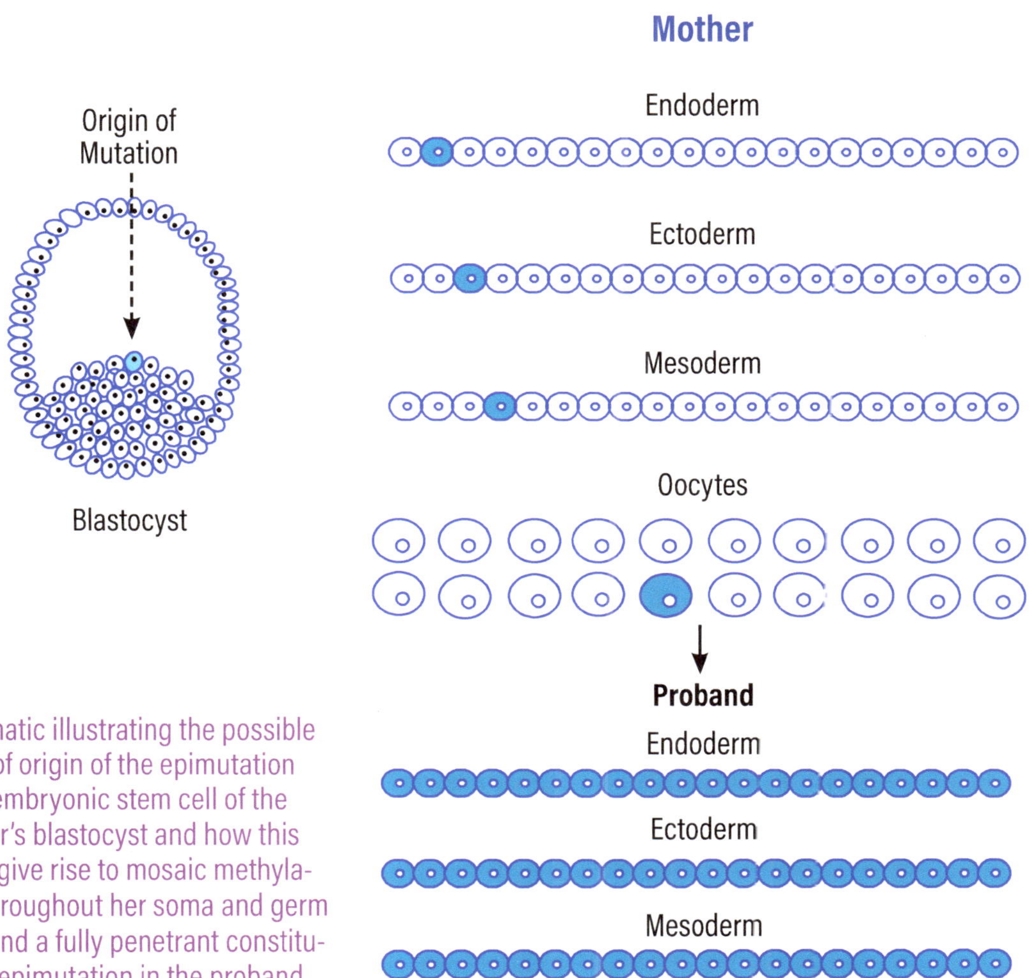

Schematic illustrating the possible point of origin of the epimutation in an embryonic stem cell of the mother's blastocyst and how this could give rise to mosaic methylation throughout her soma and germ cells and a fully penetrant constitutional epimutation in the proband.

the areas of interest in explaining tumor heterogeneity is the early losses in onco-suppressor gene activity (such as TP53, which is the custodian of genomic stability), or gains.

Long regarded as the "guardian of the genome," the functions of TP53 as a tumor suppressor have been extensively studied.[103] More than half of human cancers possess mutations in either TP53 or genes whose protein products control TP53 activity (MDM2 or CDKN2A). TP53 exerts its tumor-suppressive activities mainly by acting as a transcription factor to transactivate gene expression following genotoxic or oncogenic stress.

Under mild stress, activated TP53 induces expression of growth-arrest genes, such as CDKN1A/p21, to facilitate DNA damage repair. When encountering prolonged or catastrophic DNA damage, TP53 triggers programmed cell death (apoptosis), primarily by transactivating genes that encode pro-apoptotic proteins, such as BBC3 (PUMA) and PMAIP1 (NOXA).

As much as we love TP53, I should mention that there is a correlation between overactive TP53 and insulin resistance! This was first investigated in 2009, and several further experiments confirmed that activation of TP53 induces diabetes by increased microRNA 34, which sensitized beta cells to apoptosis and inhibited the insulin exocytosis pathway, resulting in impaired insulin secretion. This is a vicious cycle because the inflammation originally causes the upregulation of TP53 and that causes insulin resistance. Further glucotoxicity triggers DNA damage and TP53 activation, which ultimately results in beta cell death and late-onset diabetes. As such, deletion or inhibition of TP53 restores mitochondrial function and increases glucose tolerance.[104] That said, TP53 activation protects against cancer development through the same mechanism. Following DNA damage, TP53 can reprogram the cell's energy-producing strategies from glycolysis to mitochondrial respiration (or oxidative phosphorylation) in order to suppress tumor progression. It also inhibits G6PD, pyruvate dehydrogenase kinase-2, and hexokinase and indirectly controls glycolysis by regulating the mTOR and PI3K/Akt pathways:

Reactive oxygen species/Hypoxia→TP53 mutation→PTEN/IGFBP3→PI3KCA

This is why in my practice, when a tumor shows inactivating mutations in TP53, automatic strategies to inhibit PI3KCA are taken, one of which is administration (prescription) of riluzole, which is an inhibitor of NMDA binding with calcium at PI3KCA target. One interesting interaction I had many years ago at a major oncology conference was with one of the pioneers in PI3KCA gene research, when I asked him a question about NMDA and he had no clue of the answer! Later they called me the wedding crasher, as any question I asked no one knew the answer to (except myself, as I never ask a question unless I know the answer to it).

Let's go back to the TP53 story:

TP53 germline mutation (Li-Fraumeni syndrome), has 100 percent cumulative lifespan cancer incidence (of many types). There is crosstalk of MDM2, CDKN2A, and TP53 (activated oncogenes→CKDN2A→MDM2→TP53), and there are MDM2 inhibitors suggested for use after failure of immune therapies in melanomas. Also, in myelofibrosis MDM2 inhibitors are used after JAK inhibitors. In general MDM-2 inhibitors are effective when there is overexpression of MDM-2.

Genomic stability is a unique concept. I have published articles on this subject, specifically addressing the role of epigenomes in genomic stability. In summary, there are three main proteins involved in harmony to keep DNA stable: KEAP1, STEX, and USP11. In this regard, neurodegenerative disorders and cancer are very similar (two sides of the same coin) in the initiation process, where genome integrity becomes compromised by destabilization. In cancer the destabilizing effect causes proliferation, and in neurodegenerative diseases the genomic instability causes cell death. As such, all three proteins discussed are targets for both therapeutic areas.

Endogenous mutants cause genome integrity to be compromised. That said, when a tumor commits its growth to a specific colony, the genome is more stable than an otherwise nonmutated tumor. For example, in the case of CLL, clonal mutation makes the tumor more stable and predictable in its biology. Unmutated IGHV tumors have a high chance of acquiring new mutations and taking off (IGHV mutations may still benefit from CIT standard chemotherapy in CLL).

Either using any therapy that targets a specific target or using chemotherapy will cause genomic instability, and many studies have confirmed the evolution of TP53 colonies after using cytotoxic treatments. Another example is again in the CLL, where deletion 11q signifies the tumor's predictable behavior, while after chemotherapy 11q can become 17p, and fast growing. (FISH studies, as well as molecular analysis and flow cytometry assays, have confirmed such transformations.) That said, BK inhibitors are still effective and adding venetoclax does not improve response. There are arguments that zanubrutinib has a better response rate in this setting, but I disagree. I also disagree that accomplishing undetectable MRD in this setting (using venetoclax) improves overall survival.

In resting cells, the enzymatic deamination of 5-methylcytosine creates point mutations from purine mismatching at the next cell replication. This is the most common transition (C→T) and the universal clock of aging. Unrepaired deleterious mutations usually lead to cell senescence or elimination by apoptosis. Mutations accumulate in adult stem cells at a rate of approximately 20–40 annually, mainly from this process (deamination).

Here I would like to speak more about the dichotomy and confusion in relation to genomic analysis for somatic and germline mutation analysis. Germline mutation analysis obviously has been the fundamental path to discovering relationships regarding family histories of cancer and individual risk during the long decade behind us. Many of the cancer types still claimed to be not inherited are actually related to specific germline mutations. Even now, patients with AML are told there is no familial risk factor; that said, germline mutations in AML are increasingly identified (RUNX1, DDX4, CEBPA, GATA2, NPM1, and somatic alterations such as FLT3, TP53).

Furthermore, with an understanding of tumor genomics and discovering somatic mutations as targets for treatment several studies were done to compare the efficacy of a specific customized targeted drug for germline as well as somatic mutations. These somatic mutations are associated with a higher rate of presence of SNP disorders, *cytogenic abnormalities,* and germline mutations (for example, c-Kit mutation in AML is highly associated with t(8;21).

These studies have had conflicting data, mostly showing that targeted drugs for germline mutated cancer have better efficacy and response rates compared to somatically mutated genes. But in some studies the results are very comparable; for example, in mutated BRCA tumors somatic mutation has resulted in similar responses.

With germline BRCA mutated disease, a significant number of patients who carry somatic mutations have germline mutations as well. In fact, there are recommendations to conduct germline testing for patients who carry specific somatic mutations in their tumor, and this is mainly to come up with recommendations for the family members of these patients to be tested, and preventive recommendations provided to them. The reality is that for the deadliest types of cancers, such as pancreatic cancer, there is only a small percentage of patients who have any family history. That said, only 50 percent of patients live long enough to have second-line therapy.

For example, only 10 percent of patients with pancreatic cancer have a significant family history and as such the likelihood of identifying a germline mutation is under 10 percent. That signifies the role of epigenetic testing in identifying individuals at risk, as genetic germline testing fails to identify most who are at risk. That said, *mismatch repair mutation or MSI testing ought to be offered to all pancreatic cancer patients,* specifically looking at BRCA1/2, PALB2, ATM, STK11, TP53 and CDKN2A, as well as germline mutations causing pancreatitis: PRSS1, CFTR, and SPINK1. The risk of carcinogenesis increases with any of these alterations between four and 134-fold. The standard of care recommendation for patients with pancreatic cancer harboring these mutations is platinum-based agents for four cycles, then a switch to a PARP inhibitor alone or in combination with a PD-L1 inhibitor!

Recently, gene-editing technologies have been tested to edit specific mutations for people at risk.

The second area of genetic testing in relation to somatic mutation or germline mutation has to do with treatment options. Some types of cancer manifest significant actionable targets and mutations, versus some other types of cancers that do not, even in one area of oncology specialties, such as GI. You see that for example hepatic cancer or cholangiocarcinomas have about 40 percent actionable mutations revealed in their testing versus pancreatic cancer, which only shows 10 percent. The other challenge is that even in a group of patients that you see an actionable target targeting, that actionable target does not necessarily mean anything for a better outcome. For example, cyclin-dependent kinases can be seen as an actionable target targeting pancreatic cancer, and yet clinically treatment of this target would not improve patient survival at all.

That is the biggest challenge, as the promise of an actionable target will not provide any meaningful outcome advantage for the patient. That said, based on the "know your tumor" profile in pancreatic cancer, the presence of actionable targets in the molecular profiling can increase the patient's survival up to 12 months (double it).

As far as targeted radiation therapies (precision radiation), SMART (stereotactic magnetic resonance guided adaptive radiotherapy) might be used for locally advanced inoperable pancreatic cancer, or MR-guided SABR as targeted radiation for NSCLC. There are also suggestions for using low-dose radiation (precision radiation) for throat cancer. There are also admissions that there is no biochemical PFS benefit with adjuvant versus salvage radiotherapy after radical prostatectomy for prostate cancer! This comes out of ASCO post-2021, after millions and millions of people have already used radiation after prostatectomies for decades!

Myeloablative doses of radiation to the bone marrow also has shown promising results in bone marrow refractory AML (radioactive iodine CD 45 antibody). Radiolabeled Y-90 and rituximab are also being used in lymphomas. Another radiolabeled drug is lutetium 177, used for treatment of leptomeningeal disease and CNS. Y-90 TheraSpheres are also studied in HCC, with very good reports, as well as sarcomas, neuroendocrine tumors, cholangiocarcinomas, and metastatic breast CA.

The intraoperative irradiation of early stage breast cancer has also failed to make any benefit in survival. That said, I have seen some positive results from both proton therapy as well as brachytherapy for treatment of prostate cancer, and even the combination of both. The current studies confirm that brachytherapy at single dose (of 24Gy SDRT) can be as effective as 5 days of 9Gy stereotactic fraction treatments.

Another example of precision radiation is “Radioligand” therapy, aimed at DDR genes (DNA damage response genes/homologous recombination genes: BRCA1/2, ATM/ATR, CHEK2, MLH1, MRE11A, FANCA, NBN, PALB2, and RAD51C), which is sponsored by Novartis.

Also, GammaTile is a radiation system in which Cesium 131 seeds are placed post surgical resection of any metastatic disease in the brain (any lesion above 2 cm is a candidate for surgical resection and further stereotactic radiation traditionally).

The concept of combined therapy or combination therapy in general (except some rare protocols) has failed to provide meaningful quality and quantity of life for patients with several actionable targets. The reality is that because each of the technologies used in treatment of cancer has minimal efficacy the clinicians tend to think that if they combined different treatments with somewhat deficient results, they would have a general additive effect with improved survival. For example, IMpower150, a three-arm phase-three trial evaluating Tecentriq or azetolizumab plus chemotherapy, atezolizumab and Avastin plus chemotherapy, and Avastin plus chemotherapy in patients with stage IV recurrent or metastatic non-small-cell lung cancer, showed only minimal efficacy in the combination of all four drugs, with the conclusion that combining targeted antiangiogenic therapy plus chemotherapy plus immunotherapy had provided about 10 percent response rate and only the advantage of progression-free survival improvement, which was negligible.

That said, over 30 percent of patients discontinued treatment due to adverse events, which is the main challenge with addition of combined treatments. On this path mostly the toxicities add, and the efficacy doesn't. That said, in December 2018 the four-drug regimen received regulatory approval for the first-line treatment of patients with metastatic non-squamous cell carcinoma. Used in an exclusive number of patients with EGFR aberrations, this sort of approval with minimum data provides false hope to the patient who puts all their hope in the combination treatment with only additive toxicity. One other example is the combination of Xeloda with neratinib in HER2 metastatic breast CA, with no improved survival (either median PFS or OS), compared to lapatinib and Xeloda, and even that had only 5.5 months of PFS. That said, the study recommends the protocol as it shows a better "mean progression-free survival" and reduced relative risk of death, which I hope everyone knows means nothing!

Another therapy studied in squamous cell carcinoma was a PARP inhibitor that yielded a 4 percent response. In squamous cell carcinomas, generally there is no single driver gene and the tumor expresses a wide range of genetic abnormalities.

These are all examples of *pseudoscience*. The very first critique of randomized clinical trials was by Dr. Freireich at MDA (the developmental therapeutics department) in 1975. By the way, I have worked with this department—not in the 1970s, but in 2015. Although the department is designed for inventions and drug developments, it is greatly influenced by the MDA's overall political environment. I will refrain from discussing my experience because it is beyond the scope of this book, but I will reference this as the topic is about the conflict that infiltrates the scientific society, starting, as I mentioned, in the 1970s. In one heated exchange between Dr. Freireich and his colleagues, described in an article by Elihu Estey MD, JCO

27,[105] he argues against the randomization of trials with minimum results, stating that "the primary beneficiary of the clinical research must be the patient, participating in that research" (not the sponsors).

The drug companies design the trials and try to come up with a concept that has no meaningful value. As an example, for the longest time we used angiogenic blockers, before we had to admit our failure. The FDA itself had to remove the drug Avastin from the market as an adjuvant therapy for breast cancer, because the outcome for breast cancer treatment in patients who were treated with this drug was notoriously inferior to the outcome for those who were not. This is when the FDA itself was under the same oncology leadership that had approved the drug 2 years prior. Remember, over 50 percent of accelerated approvals by the FDA *lack* confirmatory evidence (aducanumab is a perfect example; that has created such a drastic argument on its efficacy, and sales over 29 billion dollars a year). Aside from the fact that the majority of these accelerated approvals are for very toxic drugs; for example, more than a quarter of drugs approved for use in cancer are severely cardiotoxic.

All angiogeneic blockers increase tumor hypoxia. A good example is cediranib, which has been suggested to increase intratumoral hypoxia and as such is believed to improve the efficacy of olaparib in prostate cancer and suggested to be used in combination with this drug (biomarker analysis from a randomized phase II study of olaparib with or without cediranib in men with metastatic castration resistant prostate cancer (Clinical Advances in Hemotology & Oncology, Vol 19, issue 4).[106] This combination added 30 days to patients' survival. The concept was so illogical that you can only question the mindset of the trial designers! I think they did not comprehend that increased tumor hypoxia although downregulates the genes involved in homologous recombination (TP53, BRCA, ATM, PALB2, CDK12,TMPRSS2), this is only *pseudoscience*, as the same drug potentiates the production of blood vessels necessary for tumor metastasis. (TMPRSS2 is also a target for testosterone transcription, as is the PSA.) One of the findings in our clinic is that when a tumor is treated with oxygen-delivering technologies such as hyperbaric oxygen or agents to inhibit hypoxia inducible factors (and as a result increase intratumoral oxygen), the tumor mutation burden increases with overexpression of all 14 HHR genes (BRCA, etc.). This is in concert with the fact that hypoxia downregulates these genes.

The term pseudoscience applies not only to the concept but also the statistics used in a trial, many of which only reflect the advantages of days in survival. One study recently published in *Clinical Genitourinary Cancer* (April 2021) showed a survival benefit of one third of a day with the protocol used (Olivia A Do, et al.).[107] I am wondering how they get published....

For small-cell lung cancer, topotecan or CVD used to be the go-to, and now the application of Lubrinectadine (which bonds with guanine residues, inhibiting the RNA transcription) is suggested along with Doxil, but this regimen is extremely toxic, in my view. Newer platinum-based agents show less toxicity; lobaplatin is an example. Largely immune checkpoint inhibitors have failed to provide meaningful results over the standard of care; specifically, the addition of CTL-4 and PD-1 inhibitors have failed to show improved survival in this setting. That said, azetoluzimab and durvalumab are now the standard of care in combination with CE and EP, respectively. Durvalumab is also approved in unresectable stage III NSCLC and also in an adjuvant setting for resectable disease; but yet again I am still waiting to see a clinical positive response in the group of patients I have seen.

In small-cell lung CA with brain metastatic disease, the 1-year survival rate is only 21 percent if treated early with Temodar (dosed at 75 mg/m2, 3 weeks on, 1 week off), and in refractory cases no one survives at the 1-year cutoff. Also Trilaciclib/Cosela (an injectable CDK inhibitor) has been promoted for reducing the bone marrow suppression along with topotecan. Other medication studied for small-cell lung cancer is tarlatamab.

I think I have discussed this issue in regards to progression-free survival as a prognostic marker for response in the past. What brings this dichotomy back into the surface seems to be the fact that not only there is no meaningful reason to check for progression-free survival (if it does not correlate with the patient's overall survival), but the fact of the matter is that many times improved progression-free survival means reduced overall survival.

This is because as the progression free-survival improves, equivalently the stem cell activation increases, which reduces overall survival; this explains why the improved progression-free survival does not translate to a better overall survival. It is as if the tumor catches up with its growth after a temporary break because of the therapy.

Also, you see that in many cases progression-free survival is less than overall survival, which again tells you that there is no correlation. For example, a combination of BRAF and MEK inhibition in patients with BRAF mutated melanoma has provided an improved progression-free survival rate of 19 percent, and yet the overall survival rate is 34 percent with the therapy, meaning that there is no correlation between the two. In general, melanomas fall into one of four genetic categories: BRAF (50 percent), RAS (25 percent), NF1 (15 percent), and triple wild (15 percent). Kit and HER2 are rare. Therapies such as lifileucel, a tumor-infiltrating lymphocyte therapy, are under development.

The network of genetic crosstalk has just been discovered in recent years. The old school of thought of finding an actionable target and blocking it is outdated, now that we know that parallel growth pathways in the tumor have significant feedback, loops, and crosstalk. For example, the EGFR family of targets have significant interaction with their phosphoinositide 3-kinases or PI3K pathway. The heterodimerization of HER2, HER3 molecule is a key amplifier of PI3K/AKT signaling, in addition to being a trigger for other signaling pathways such as the MAPK pathway. (This is why patients with EGFR+ NSCLC at our clinic receive PI3KCA inhibitors, along with EGFR TKi.)

EGFR/FGF→+PI3KCA→+AKT→(-)FOXO1→(-)TSC→
(-)mTOR→(-)GSK→(-)Wnt

As you see, the PI3K is activated by growth factors, and further, the cascade will inhibit the TSC and FOXO1 (which are both tumor suppressor genes), and mTOR is inhibited by the TSC. Activation of mTOR will decrease the inhibitory effect of GSK and increase the Wnt (GSK regulates the Wnt). mTOR inhibitors are certainly useful in TSC1 mutations.

Copanlisib and idelalisib (a dual PI3K and CK1 inhibitor) and umbralisib (a specific gamma isoform inhibitor and Wnt pathway inhibitor) are PIK3CA inhibitors, and capivasertib is an AKT inhibitor (ineffective in breast CA), and there are AKT/HER3 combo inhibitors undergoing study. The TP53 mutation is highly associated with HER2 breast CA. Copanlisib is also suggested in Waldenstrom's macroglobulinemia, along with rituximab. Other PI3K inhibitors are buparlisib (studied in SCC) and paxalisib (studied in glioblastomas). Copanlisib is used with Faslodex and Verzenio. Also suggested in lymphomas along with rituximab, or the newer version of it, is naratuximab. Umbralisib is also approved for follicular or marginal zone lymphoma as a third-line therapy (After R-CHOP and BK inhibitors). Follicular lymphomas have translocation (4:18) in 90 percent of cases. Marginal zone lymphomas are of three kinds: MALT syndrome or mucosal, which is the most common type; extranodal (bone marrow or liver), and splenic. The splenic form is associated with hepatitis C infection.

In lung cancer, inhibition of the HER2 molecule, or the fusion of NRG1 inhibition, has a significant role in the crosstalk with the P38 section or regulatory subunits of PI3K protein. The drugs afatinib and Seribantumab are able to reduce or inhibit the NRG1 fusions and thus the HER2, HER3 tyrosine kinase receptor pair or heterodimerization. Patritumab deruxtecan is a anti-HER3 medication used in combination with irinotecan on EGFR+ lung cancer, as 80 percent of these tumors carry overexpressed HER3. Mutations, however, are rare (about 2 percent), after CDKN2A (6 percent) and MET amplifications (8 percent).

The same is true for using chemotherapy for stage II lung cancer, which is questionable, and depends on the tumor's cyclindependent kinase activity, but the standard of care offers adjuvant chemotherapy for all tumors larger than 4 cm! For stage II nivolumab is offered with chemotherapy to accomplish a 77 percent 2-year survival rate.

Overall, addition of immune therapies seems to improve survival in KRAS+ lung cancers (30 percent), but not EGFR/ALK+ tumors. These are generally individuals who smoke. That said, patients who have overexpression of amplifications of the EGFR are usually nonsmokers, as well as female. Example of ALK inhibitors include alectinib, brigatinib and lorlatinib. ALK+ lung cancers entail about 5 percent of total cases. ALK is also positive in about 10 percent of neuroblastoma cases at the clonal level.

A rare form of EGFR mutation is exon 20 insertion, which until recently there were no drugs for; however, mobocertinib and poziotinib are now approved as TKI and amivantamab as antibody. Other drugs for HER2+ EGFR are pyrotinib, apatinib, and neratinib.

Nimotuzumab is a monoclonal antibody that has been effectively used in patients with locally advanced NSCLC with positive EGFR mutation, providing a survival benefit of 5 years of about 58 percent, if concurrently used with chemo and radiation. It is also suggested as a trial medication in addition to Temodar to treat patients with glioblastoma. Not knowing the

EGFR positivity of glioblastoma, the medication is tested blindly, yet with a promise that EGFR- tumors still respond to this medication (80 percent of GBM carries EGFR amplifications identified in the signatures of liquid biopsies). Although standard EGFR inhibitors pass to some extent through the blood–brain barrier and I use them in glioblastoma very effectively, recent studies have looked into better penetration of the blood–brain barrier by a new drug called ERAS-801. I personally have doubts about the mechanism of action of this drug, knowing that it may block the downstreams of the pathway, resulting in quick resistance.

The same concept was tested in squamous cell carcinoma of the head and neck through a phase II trial in 2018, with an additional 4 months of OS when added to chemo and radiation. This concept is pretty much contrary to precision oncology, but it signifies an important misconception about using targeted drugs. That said, tipifarnib (a fernyltransferase inhibitor) is now approved for HRAS+ SCC of the head and neck if the cDNA has at least 20 percent MAF. This 20 percent is not a random designation; rather, it comes from the fact that clonal expansions need at least 20 percent mutated variant frequencies; anything less would be considered a "truncal," or branch, mutation (subclonal).

In general, afatinib is the medication of choice in metastatic squamous cell carcinomas, head and neck cancers, or lung cancers (with or without necitumemab), as EGFR is overexpressed in 90 percent of head and neck cancers (as is PIK3CA). As patients receive radiation, EGFR becomes overexpressed and translocates to the nucleus, where it acts as a transcriptional factor. The connection between HRAS and PIK3CA is interesting, as the helicase domain of PIK3CA mutant must bind to RAS to cause the squamous cell cancer (the RAS family has three subtypes: KRAS is frequently found in adenocarcinomas, HRAS in SCC, and NRAS in leukemias and melanomas). RAS+ mutations generally overexpress low tumor mutational burden. Stages I and II of head and neck cancers generally are treated only with surgery or radiation alone with a 70–90 percent cure rate.

The concept of using epigenetic drugs started with hematological cancer and one of the first indications was myelo dysplastic syndrome, or MDS, for the demethylating drugs. Later on, the FDA approved a combination of decitabine as a demethylating drug with other medications for high-risk MDS, such as cedazuridine. Also, the combination of demethylating and cytidine deiminase inhibitors was suggested by Epigenetics Dream Team led by Stephen Baylin from John Hopkins as he revealed that the combination of two classes of epigenetic modifiers would have better results. Now the classes of epigenetic therapies have evolved from HDACs and DNMT and can include ubiquitination and SUMOylation inhibitors, and others.

One of the mechanisms of resistance to immune therapies is thought to be the un-transcribed NKG2DL, and as such the use of HDAC inhibitors is suggested to overcome such resistance. As such, entinostat is combined with immune therapies in solid tumors.[108]

But with all that effort solid tumors were still refractory to epigenetics treatments alone. There are several reasons for this failure. The first is epigenetic modifiers, as synthetic drugs are nonselective. This meaning that, for example, a DNA demethylator of an oncosuppressor gene also demethylates oncopromoter genes, and as such the effects are very minimal.

Second, the potency of these drugs on specific CPG islands among oncosuppressor genes is not enough. There are studies confirming that natural products can have significantly higher potency on specific CPG targets. For example, EGCG has shown to be several times more potent compared to specific demethylators on specific sites. Furthermore, demathylators such as decitabine can be carcinogenic because they can activate the S100B gene and cause cancer. That said, epigenetic therapies with DNA methylation inhibition (including decitabine) have shown immune modulatory activity with increased recognition of cancer STEM cells by the immune system (when studied in gliomas). "Hypomethylating agents have the potential to enhance glioma cell recognition and subsequent destruction by immune cells, regardless of their differentiation status."[109]

It is important to note that the sequence of progression from mild dysplastic syndrome to progressive MDSN and further AML could be also regulated through epigenetics. In the mild form of MDS, where there is no treatment necessary and watchful waiting is advised, the effort is to reduce the need for transfusion and to improve the quality of life. A medication called Imetelstat is able to target the overactive telomers in this disease. This drug has also been investigated in other hematological cancers, including AML. "Imetelstat demonstrates efficacy in a significant proportion of AML models. Robust responses to Imetelstat were associated with favorable cytogenetic risk groups and mutations in pathways controlling DNA damage."[110]

Imetelstat as a telomerase inhibitor is also applied for MDS and myelofibrosis, after JAK inhibitors (ruxolitinib and federatinib). Finally, CD47 antibodies are used to treat the MDS. CD47 is also called the "do not eat me" antibody. Another target is GNAS.

Furthermore, a driver gene mutation in SF3B1 (as well as co-mutations on RUNX1 and ASXL1) can determine the fate of the disease, from mild dysplastic syndrome or mild disease into progressive disease. Truncated ASXL1 has been noted to increase the protein BAP1 and its recruitment to chromatin, as a cause for leukemias in general. It is also important to note that the traditional use of chemotherapy in a variety of hematological cancers has shown significantly lower outcome success when compared to more advanced combination of therapies, including epigenetic treatments, and as such the concept of using second- and third-line treatments has shifted to using epigenetic treatments as the first-line treatment. The same is true of using intrathecal chemotherapy for both prophylaxis and treatment of CNS disease in hematological cancer, with absolutely no merit, causing grave morbidity and mortality, to the point that the whole concept is now disputed.[111]

As we know, the process of progression of disease and failure to accomplish achievement has to do with their epigenetic appearances and as such it makes total sense to start a patient with epigenetic treatment as the backbone of any other additional treatment.

For example, in the case of diffuse large B-cell lymphoma we know that the first-line treatment of RCHOP and rituximab has shown about 30 percent to 50 percent failure rate. This is even true for follicular lymphomas (if there is increased microglobulin2, older patient age, or bone marrow involvement, there is a 2–3 percent conversion to large B-cell lymphomas). That said, a combination of ibrutinib, Afinitor, and pomalidomide has proven to be effective for relapsed diffuse B-cell lymphomas, or Richter's transformation. Ibrutinib's mechanism of action is beyond the BK target and it includes inhibition of NF-kappaB. As such, in my practice this drug is commonly combined with other therapeutic regimens, including Afinitor, for different types of cancer, including RCC and lymphomas. This is a unique approach and it is exclusive to Ibrutinib and not other medications in the same class of BK inhibitors. The combination of Ibrutinib with epigenetic therapies enhances the efficacy of this medication and sustains its effect.

Follicular lymphomas, like other indolent lymphomas (marginal zone lymphomas) are always influenced by both epigenetic and genetic dysfunction. As we know, the t(14;18) rearrangements are necessary to develop the disease but further TP53 mutations, macroglobulin mutations, and changes in the microenvironment also happen. In follicular and specifically marginal zone lymphomas the mTOR inhibitors have found their way in therapies. I personally have treated these patients using epigenetic therapies very successfully when tracking the markers (Beta 2 microglobulin, LDH, and IL-2 soluble receptor) showing a response in less than 2 weeks. One rationale on the magnitude of the response is related to the epigenetic influence on the mTOR pathway. (This concept was validated in March of 2021 by Yang et al. and published in *Cell*.[112]) Even in cases of triple-hit lymphoma, where BCL2, 6, IGH, and c-Myc are mutated (and in both activated large B-cell lymphomas versus germinal center B-cell subtypes), using epigenetic therapies has shown significant improved outcome, in my experience, when the standard of care would be ICE regimen followed by stem cell transplant, polatuzumab, tafasitamab, loncastuximab, bispecific antibodies, or CAR-T therapy. My patients received none of them, as they were not needed.

The entire concept of curable versus incurable lymphomas (as suggested by Smith MD in Clinical Advances in Hematology & Oncology, Oct, 2021)[113] to me seems irrelevant and honestly laughable. The concept promoted is that when a lymphoma is aggressive it is curable and when it is not, it is incurable. This concept reflects the reductionist thinking that defines cancer treatment, as tumor and cure are defined when there is no measurable disease burden in scans. This is *not* what I call advanced medicine to be published in advances in a hematology journal! Now, we know that both aggressive and indolent lymphomas have epigenetic disruptions which are not addressed by cytotoxic therapies and measurable disease in scans has nothing to do with cure.

On the subject of using BTK inhibitors, although I have not experienced failure, theoretically these agents can be used in mantle cell lymphoma, along with bortezomib, venetoclax, and lenalidomide alone or in combination with rituximab, when BTK fails. Other BTK inhibitors include acalabrutinib and zanubrutinib, and the most recent one, pirtobrutinib, a noncovalent irreversible inhibitor, is in the pipeline.

Also, cirmtuzumab (a monoclonal antibody against ROR1) is being studied along with ibrutinib. Acalabrutinib has same efficacy as ibrutinib, just with less frequently reported afib.

Ibrutinib is being studied in combination with rituximab in both MCL and CLL, and also in the LOXO trial on B-cell lymphomas. In most effective cases (Hodgkin's lymphoma), the risk of infertility and secondary malignancies are still a major concern for traditional chemotherapies. The main mechanism of action for anthracyclines and TOP2 inhibitors to cause secondary malignancies is through inhibition of TP53, and a possible cause for secondary MDS/AML. As such, there have been studies showing two cycles of ABVD are as effective as four cycles. It is also true that the same patients are at risk of breast cancer when radiated to the mediastinum (especially if they have BRCA mutations, coronary atherosclerosis, or mental decline). The risk of breast cancer is about as high as a carrier for BRCA mutation in survivors of Hodgkin's lymphoma after breast bed radiation, and if affected, the mortality is higher. Just to add to the problem, recent studies have shown there is 100 percent increase in risk of heart disease after left-breast radiation in women. In my practice, patients with breast cancer are not routinely referred for radiation. To my recollection only proton therapy has been advised in certain cases with refractory or recurrent disease.

I explained earlier that in CLL with TP53 mutations we already know that chemotherapies in general are ineffective. With common sense, one can reason the ineffectiveness of such agents in other types of cancer when TP53 is mutated. That said, etoposide is suggested in patients with TP53 mutations when osimertinib fails. One mechanism in my view is the fact that osimertinib causes the lung cancer cell phenotype to become closer to small-cell lung cancer, and as such the etoposide is more effective in this scenario.

Therefore, the investigators have turned to other technologies and treatments, such as post-transcriptional modification medications like as Avadomine, as well as the use defactinib and trastuzumab for first-line treatment, which is a cytolytic CD-19 targeted monoclonal antibody. We also know that using decitabine and prednisone has been associated with about 93 percent complete response in cases of ALL. In the case of CML, drugs such as asciminib (with or without bosutinib) have been used to provide meaningful response in progressive disease when the patient cannot accomplish major molecular response, defined by BCR-AB levels of less than 0.01

percent. It is known that in the case of CML the progression of disease has to do with the molecular amplification of T315I (which confers resistance to the TKIs, including dasatinib). Ponatinib, although newer, is very toxic (it is used in ALL/CML), but in combination with asciminib it overcomes the resistance to TKIs. Other agents being studied in CML include tipifarnib and pan Aurora kinase inhibitors (danusertib). Off-label drugs such as Actos have also been studied in this setting with nilotinib. Synribol and glasdegib have also been used. The main concern with TKI use in CML is cardiovascular events. Overall there is 30–40 percent treatment free remission reported in CML.

Bosutinib has also been studied in breast cancer due to its inhibition of SRC.

In ALL, using hyper-CVAD alternating with Bcl-2 inhibitors, or if chromosome Philadelphia is positive with TKI2 (newest olverembatinib) is suggested.

When a patient accomplishes remission again it is notable that epigenetic drugs are very effective in maintaining that remission. For example, in the case of diffuse large B-cell informal we use epigenetic treatments very effectively in the clinic to avoid further recurrence of disease and progression.

There is significant synergistic effect with the drug Ibrutinib, which is an important kinase inhibitor and epigenetic modifier. There is a known mechanism of resistance to Ibrutinib caused by activation of the gene PLCG2, TRAIL, induction by 8p deletion, ITPKB mutations, and CARD11 gain of function or acquisition of point mutations at the BTK SH2 binding site. There is also a synergistic effect seen using antihistamines in patients with lymphoma as well as CLL. At times, antibodies against B-cell activating factor receptors are used to inhibit this:

The doubling time is no longer an indication for severity of disease, as it can imply using chemotherapy where it causes evolution in the tumor. In general there are three classes of drugs approved for CLL: PI3K inhibitors like idelalisib, duvelisib, BCL2, and BTK inhibitors), as well as standard chemotherapy with cytarabine, Cytoxan, and rituximab [FCR]. Ublituximab (a CD20 antibody) was recently studied in combination with Ibrutinib with superior results, even in a group of 17q and TP53 patients. A triplet protocol consisting of ublituximab, venetoclax, and umbralisib has been tried.

Tazemetostat is another example of a targeted therapy as an HDACi for follicular lymphomas and sarcomas carrying EZH2 mutations (or BCL2 mutations). EZH2 mutations are now reported in endometrial cancers, sarcomas, and prostate cancers. Peroxiredoxin 6 is an intermediate gene. EZH2 is involved with all steps of cancer growth and metastasis. EZH2 inhibitors can regulate the expression of SWI/SWF chromatin remodeling in sarcomas and have been recently studied. Other off-label drugs such as "gold" can similarly inhibit the SWI and can be considered in sarcomas.

The EZH2 inhibition reduces the breast cancer's ability to convert to breast cancer stem cells in skeletal metastasis. The concept of metastatic disease as the terminal end point for distant organ disease is now challenged, as we now know that the breast metastatic lesions in the bone are able to re-disseminate after EZH2 activation to stem cells.

The combination of EZH1 and 2 is used to treat relapsed non-Hodgkin's lymphoma (valemetostat), where other options could include transplantation after a positive PET scan (the Italian method). What all recent studies have confirmed is the *lack* of benefit from radiation in Hodgkin's lymphoma (specifically when the PET is negative post ABVD). It is well documented that an interim PET scan can predict response. The best strategy in my opinion is to track the biomarkers for Hodgkin's lymphoma (such as thymus and activation regulated chemokine) to monitor response.

In endometrial cancer, a new study showed using a levonorgestrel IUD with or without weight loss/metformin was able to cause complete remission at stage I for 60 percent of cases.[114] Other potential therapies for other kinds of solid tumors could similarly target the TP53 and CDK (sequential CDK inhibitors with mTOR inhibitors).

The synergism also is obvious with the drug Ibrutinib. The HDAC inhibitors have been used very effectively in newly diagnosed AML patients with activating FLT3 mutations in combination with chemotherapy and further used as a maintenance treatment, such as the drug midostaurin. Giltritinib, quizartinib, and crenolanib are also being studied for both maintenance (alone) and first line (along with cytarabine and cladrabine or venetoclax). In my practice, Ibrutinib is used at a low dose, based on studies by Chen et al.),[115] and Advani, which showed a low dose has the same BTK occupancy as a high dose.

Midostaurine has also been studied as a protein C kinase inhibitor in solid tumors in general. I am not a fan due to the severe leukopenia it can cause; that said, I see a role for it in rectal CA as a neoadjuvant therapy to radiation, along with heat shock protein inhibitors.

Midoustaurine is also suggested along with other c-Kit inhibitors in CML (Sutinib, Nexavar, dasatinib, nilotinib), none with survival benefit in Gleevec-resistant cases (ATP binding pocket mutations).

Avapritinib and ripretinib are being studied in these circumstances. One would suggest that these medications can also have a role in AML.

Also, in AML azacitidine has been used with venetoclax or navitoclax, which inhibits BCL2 very effectively and has proven to be significantly better than one drug alone. The mechanism of resistance to venetoclax includes homozygous deletion of CDKN2A and mutations in BTG1. Knowing that venetoclax causes resistance, I wonder how it is recommended that a patient can take this medication "indefinitely"?? (The therapies basically will be withheld for only 14 days if a bone marrow biopsy shows less than 5 percent blasts.)

The combination of venetoclax and Ibrutinib has also been shown to be effective in patients with CLL. The oral form of azacitidine, called CC 486, has been used as a maintenance treatment for patients with AML in remission. In the case of MDS as well, azacitidine has been used along with CD47 monoclonal antibodies such as magrolimab. Lenalidomide, Gazyva (a CD20 antibody, generally used in CLL), Imetelstat, and cedazuridine are other medications used, and are generally reserved for patients who have a less than 8-week transfusion interval, with the aim of reducing transfusion need to 16 weeks (decitabine and cedazuridine are approved for MDS and CMML). Magrolimab targets the macrophages' "do not eat me" signal. There are also new drugs that target tumor-associated macrophages.

It is important to differentiate MDS and myelofibrosis. In MF, which happens generally in patients with autoimmune disorders, there is no dysplasia reported in the megakaryocytes, the patient presents with similar constitutional syndrome and thrombocytopenia, and the genomic mutation assays can be similar (JAK2, ASXL1, EZH2, IDH1/2). The initial CBC with low platelets always accompanies the smear that shows atypia, leukoerythroblastosis, and treatment options are ruxolitinib as well as older medications such as hydroxycarbamide. Fedratinib and pacritinib are newer medications that are used for MDS, as they also inhibit FLT3. Unlike CML, there is not a single mutation that can be targeted in MPN. Transformations to positive TP53 are common after using selective therapies. (Myeloproliferative neoplasms are categorized as difficult to treat types of leukemias where survival for 5 years is less than 40 percent.)

Furthermore, I would like to discuss the importance of germline testing for patients with microsatellite instability. One of the main areas of cancer research focuses around identifying the risk in family members of patients who have microsatellite instability high or MSIH tumors. Next generation sequencing platforms have been used to identify these patients with Lynch syndrome in about 16 percent of MSI high and about 1.9 percent of patients with indeterminates or microsatellite stable tumors. This means that the microsatellite instability has been a very good indicator that many of these patients, especially the younger ones, and also African American patients, have a higher risk of identification of germline mutations, which increases from 2 percent up to 20 percent in certain populations. For example, in breast CA, the presence of TP53 mutations are highly seen in young adults (less than 30 years old) with a positive family history. TP53 concurrent mutations include EGFR and Rb. African American patients also have seen higher incidence of germline mutations in CHEK2 and PALB2.

The only medication approved for TP53 mutation in MDS is eprenetapopt (which is effective when the blast number in BM is less than 30 percent). Adavosertib could also be considered in these circumstances as the G2/M phase is a checkpoint for TP53 mutations (this is a similarity of endometrial and ovarian serous carcinomas and AML)!

In general, approximately 20 percent to 30 percent of colorectal cancers are MSI type, and about 3 percent correspond to Lynch syndrome. The patients who have this condition obviously are thought to have a disease that could be targeted with immune therapies more effectively. (Lenvatinib has been used along with checkpoint inhibitors in colorectal CA/MSI deficient, melanoma, papillary thyroid CA, mesothelioma, and RCC.) Also, it is speculated that epigenetic therapies could be much more effective in this specific population.

Other therapies approved for RCC include nivolumab and cabozantinib as first-line therapies (instead of Sutent), and Azetolizumab and Avastin as first-line therapies (instead of nexavar) in HCC. Donafenib has also been studied as a superior drug to sorafenib. In HCC the subcategory of the disease defines the response to different therapies; for example, in non-alcoholic fatty liver disease–caused HCC, the checkpoint inhibitors actually *increase* the fibrosis and cause a higher incidence of HCC.[115]

Tivozanib is also approved for RCC as a VEGF inhibitor, along with levantinib. (Levantinib is also used in ovarian and endometrial CAs.)

There has been no efficacy reported from immune modulating drugs, or targeted drugs such as Sutinib or nexavar in mesotheliomas. That said, there is an additional 5-month benefit from PD-L1 inhibitors (11 months versus 6 months OS reported from chemotherapies). This is very similar to studies in small-cell lung CA.

Role of ten-eleven translocation enzymes: TET is a demethylator and counteracts DNMT1. Some actually consider this enzyme a tumor suppressor gene. Both vitamin A and vitamin C increase the levels of TET1 or its function.

The discovery of TET enzymes was one of the biggest recent discoveries in epigenetics. They provided a mechanistic basis for a mostly hypothetical pathway, active DNA demethylation. The enzymes are named for a common translocation in cancers. A translocation can occur between chromosomes 10 and 11, creating an MLL-TET1 fusion protein, a common feature for AML. TET is mutated in many types of solid tumors as well. It is responsible for demethylation of 5-methylcytosine (5mC). 5mC is first converted to 5-hydroxymethylcytosine (5hmC), then 5-formylcytosine (5fC), then 5-carboxylcytosine (5caC), each by TET1-3. 5hmC is seen in the brain at high concentrations.

It is important to know that under hypoxic conditions, TET/HIF form a combo molecule that increases tumor aggressiveness and proliferation. *As such, the activation of TET only should be suggested after the HIF is inhibited.*

Hypoxia, Epithelial-Mesenchymal Transition, and TET-Mediated Epigenetic Changes[117];

Hypoxia→HIF→TET→HIF

HIF inhibition reduces the mTOR resistance through TSC1/2 and REDD1. Recent research has shown promise in using TET2 as a target to treat T- ALL in children.

In the case of hematological cancers, the use of genomic labs can assist with identification of actionable targets in the blood. This test can identify alterations, such as TP53, IDH1, TET1 and 2, and DNMT1. For example, in the case of AML treatments that block DNMT1 (decitabine) have been implicated, along with use of HDACi, although the presence of DNMT1 mutation has not been suggested to be prognostic. (The presence of TP53 mutation has better a prognosis for response to therapy and the presence of *STAG2/IDH2/ASXL1* with low response rates. The newer form of decitabine, called Guanacitabine (the addition of deoxyguanosine to decitabine), causes gradual release of decitabine and lower toxicities).

In AML, there is also SNDX-5613 in trials to treat it epigenetically. This drug targets DOT1L and H3K79 and is effective in MLL rearranged/+ (through MLL menin interaction) AML.

When it comes to MRD in AML, 30 percent of patients with MRD relapse, and 30 percent without MRD don't, so the concept of MRD is now under scrutiny, at least in my view. A stem cell transplant on someone with positive MRD is basically stupid, in my view. What I generally recommend to patients with positive MRD is to use epigenetic therapies (decitabine even in an oral form, is effective). MRD is also a goal of therapy for CML and CLL. The best validated test for MRD is PCR, but NGS is suggested as equally validated as bone marrow biopsy (Oncology Times, May 2021),[118] although I disagree. I still believe that the PCR test is superior to NGS and NGS is not equivalent to the negative bone marrow negativity. That said, whole genome sequencing is better than traditional cytogenic analysis.

There is recent improvement in our understanding of the relationship between genes and epigenes. We now know that epigenetic enzymes impact one another. For example, DNMT1 and HDAC can each interchangeably induce the other. HAT counteracts HDAC (so does Sirtulins). ERK and KRAS both activate DNMT1.

HDAC→ERK→ DNMT1

I would like to discuss the role of methylation markers as diagnostic tools for diagnosis. For example, methylation of GSTP1 is now equivalent to tissue biopsy for diagnosis of prostate cancer. Also in cases where the biopsy is negative, the methylation of fibroblasts is diagnostic for prostate cancer. These fibroblasts are triggered by the tumor present in the prostate and surrounding the tumor form the microenvironment. It is also well known that epigenetics play a great role in prostate cancer regulation. There is the ChIA-PET technique investigated on the role of RNA polymerase and enhancers epistasis for chromatin interactions to study the transcriptional regulation of prostate cancer. Glucocorticoid receptors also have been studied thoroughly in prostate cancer, to suggest the use of their inhibitors (Korlym) or RU 486 in the setting of advanced prostate cancer.

DOCK1:

DOCK1 is a common mutation for all identical twins with breast CA. DOCK1 is related to RAS mutation and tumor glutamine consumption. (There have been glutaminase [GLS1] inhibitors discovered that will be studied soon in NSCLC with STK11 mutations and loss of KEAP1 function).

There is a new KEAP1 (taleglenastat/Keapsake) currently under approval process for NSCLC.

Glutamine is unfortunately recommended by many physicians to cancer patients, and although it is helpful in regeneration of the gut, and also is helpful in management of sickle cell anemia by increasing generation of NAD and arginine, it is notorious in cancer therapy, if used.

The other explanation for carcinogenesis in identical twins, when one of the pair is impacted, is the possibility of developmental mutations before twinning, although this theory in my opinion is senseless. I do agree that developmental mutations play a big role in childhood cancer, and in certain types of hematological cancer later on, such as MDS and AML. In MDS the splicing gene mutations (SF3B1, SRSF2, U2AF1) can be dominant genomic features that happen in fetal stages. The less there are genomic mutations, the greater the role of epigenetics in carcinogenesis.

I also agree that some stem cell markers become aberrant early on. This includes NOTCH1 in GI tumors (explaining the high prevalence of GI malignancies in young individuals, in this century) and other somatic mutations which rise during early development and continue to persist in the progeny of the originating cell, but I also recognize that genomic instability is *not* enough to cause cancer, and the perfect example of this is the placenta tissue, which has the highest degree of mutation burden, and yet no tumor.

As such, the role of epigenetic regulation in controlling the cascade of endocrine machinery and transcription is important.

DOCK1 is related to production of interleukin 11, and is seen in melanomas, RCC, lung, and other types of cancer. Its presence is associated with the RAL-GEF pathway, PI3K activation, and microRNA 486, detectable in the blood. It also activates NF-kappaB, Snail. MicroRNA-486-5p and microRNA-486-3p are multifaceted pleiotropic mediators in oncological and non-oncological conditions. KRAS co-mutations include TP53, LKB1, and STK11. Another co-mutation for KRAS is KEAP1, which is a negative regulator of NRF2 and is involved with glutamine metabolism. (There have been glutaminase [GLS1] inhibitors discovered that will be studied soon in NSCLC with STK11 mutations, and loss of KEAP1 function.) DOCK1 overexpression is also prognostic in AML.[119]

DOCK1 is *reduced* by vitamin E, alcohol, Dimethylhydrazine, atrazine, and Aflatoxins, and *increased* by Acrylamide, benzopyrene, bisulfide, folic acid, and bisphenols. Lead, Tylenol, and Valproic acid have mixed reactions.

EGFR concurrent mutations include TP53 (most common in 69 percent of cases), RB1, and CTNNB1. (DOCK1 [dedicator of cyto-kinesis 1] Rat Genome Database [mcw.edu]).

Lessons to be learned from CLL:

Patients with CLL who carry 13q14 deletion, as a first step of genetic abnormalities (60 percent of CLL cases) have TP53 mutations and a predisposition for cancer. We earlier discussed the role of microRNA 15 and 16 in relation to TP53 mutation. These safeguards lose their ability to control apoptosis and Bcl-2 will increase. These patients do not benefit from chemotherapy. (If we broadly apply this concept to solid tumors, over 45 percent of tumors should not receive chemotherapy due to mutations of TP53 or inactivation of the gene due to deletion.) We see this so often in every trial, from stage II lung cancer to stage I endometrial CA, and breast CAs at all stages. For example, although 85 percent of stage I endometrial CAs have myometrium invasion, and they are really not stage I, adjuvant chemotherapy does *not* improve survival. Adavosertib as a WEE1 inhibitor is suggested for endometrial, pancreatic, colon, and ovarian cancers with improved survival in RAS mutated disease.

These patients would benefit from ibrutinib/venetoclax, for example in CLL, as their response is independent of TP53 status. Other variants of mutations include ATM (11q23) and NOTCH1 (trisomy 12). If chemotherapy is used in patients with CLL, clonal evolution happens five times more often (selective advantage) than in a non-chemotherapy group. In other words, the patients who are not on chemo have only a 10 percent risk of clonal evolution (either linear or branching evolutions), and as a result drug resistance compared to 52 percent on chemotherapy. In patients with breast cancer, after using hormonal blockade we see 40 percent acquire resistance to HT. The mechanism of resistance is thought to be through FGF signaling.

Epigenetic immune modifications: There are studies showing that post exposure to live vaccines (polio, measles, BCG, or beta glucan), the immune system recognizes infectious particles and cancer cells over the long term with general alertness, through epigenetic influence (the memory of immune system depends on IL-7). One example is using the cholera vaccine before colectomy in colon CA, or the flu vaccine before mastectomy (Gauthier Bouche, et al., Journal of Clinical Oncology).[120]

Patients vaccinated for these elements have had a better outcome with Covid-19, for example. The story of Covid-19 goes beyond this publication, but certainly, since I hold several patents in treatment of Covid-19, those interested can search my publications in this regard. One key replication gene in Covid-19 is PI3KCA, and as such epigenetic therapies are effective in treatment of this disease. Other immune modulating drugs such as JAK inhibitors and rituximab (and even imatinib) also have been studied. Beyond my own publications, Michael Foote published an interesting article in JCO, December 2021,[121] showing that two antimetabolites and three PI3KCA inhibitors were effective against Covid-19 by reducing the ACE-2. He only missed the fact that the mechanism of efficacy is not only through ACE but rather direct PI3K inhibition.

The protection caused by other vaccinations in cancer is through macrophages and not lymphocytes (increased H3K27ac and reduced 3met, increased H3K43me, and chromatin accessibility, induced TNF and IL-6). As a result the glycolysis, glutaminolysis, and cholesterol synthesis increase. Interestingly. CTLA-4 inhibitors inhibit this process (H3K27ac, H3K9 met).

On the other hand, the cancer vaccines that are developed using adoptive cell therapies rely on training the lymphocytes to recognize and attack cancer cells. GVAX and OpenVax are good examples as synthetic double-stranded RNA, used in several patients in a small trial to induce de novo immunity. One study suggested improved effects of vaccines used in treatment of melanoma when combined with GCSF. Other studies show HPV 16+ cancers benefit from intratumoral or systemic application of attenuated live viruses presenting nononcogenic HPV neoantigens (E6/7 fusion proteins), or glioma vaccine (TEM-GBM), or HER2 vaccines in patients with HER2 breast CA. Another means of delivering oncolytic adenoviruses is mesenchymal stem cells. Adoptive cell transfer of tumor-infiltrating lymphocytes also has failed to provide a meaningful response in patients who have failed other therapies.

Decitabine has been able to restore the immune system's (CD8+) ability to recognize pathogens.

Chapter Nine:

Tumor Metastasis

Basically, patients die of metastatic disease. Nearly half of patients with GI malignancies present with metastatic disease. The only option at that point is palliative care, and only 15 percent of patients on average receive that.

I want to speak about the relationship of mitochondrial DNA to the process of tumor spread or distant organ metastases. The fact of the matter is that they have mutations that occur in the mitochondrial DNA and play a major role in dissemination of the tumor cells. As such, it is important to consider the importance of intra-tumoral hypoxia and the role of epigenetics in this regard. Since metastasis remains the major cause of cancer-related deaths, it is important to identify possible therapeutic options that target the mitochondrial DNA influence on this process.

It is important to know that when a tumor disseminates the phenotypic characterization of the cells at distant organs mimics the primary tumor at that site. For example, tumor cells in the liver, when metastasized from the breast, have the characteristics of the primary liver tumor. One theory is that the stroma/microenvironment influences the tumor cells to typically present a certain phenotype. Furthermore, there is reported enrichment of ASXL1 amplifications and PTEN deletions in brain metastasis, an enrichment of DNMT3A mutations in bone metastasis, NOTCH1 mutations in skin metastasis, and enrichment of KRAS, and co-mutation of KEAP1, STK11, and EGFR mutations in lung metastasis (drivers of site-specific mutation).

In the case of breast cancer, cells that spread to the bones or the liver are more ER+ compared to those that spread to the lungs and the brain, and HER2 is more positive in brain metastatic disease.

One other important factor in the determination of cancer cells to spread is the original microenvironment. For example, the stiffer the original environment is, the more likely the cells will spread. One question that always came to my mind was why some patients with breast cancer have very large tumors with essentially no metastatic disease and some spread very quickly. One obvious reason is certainly the tendency of the tumor to build blood vessels and further disseminate, which will be influenced by epigenetic factors (both production of blood vessels as well as colonization of cells in distant organs), but beyond that the matrix fibrin content and FGF, and as such one of my initial approaches was to inhibit the FGF by antifibrotic agents. These agents have recently been proposed by scientists at the University of Arizona, who published data in *Oncology Times*, in October 2021, showing that drugs used to treat pulmonary fibrosis (pirfenidone) can reduce the risk of bone-metastatic disease in breast cancer patients.[122] Pirfenidone (PFD, Shionogi & Co., Ltd., Osaka, Japan) is a pyridine compound with therapeutic potential for idiopathic pulmonary fibrosis. In previous studies, PFD has been shown to inhibit fibrosis and/or desmoplasia in vivo, as well as proliferation and/or activation of fibroblasts in vitro in various organs and cell lines. In addition, Jin et al. analyzed mechanisms of the lung fibroblast-dependent anti-fibrotic effects of PFD and showed that TGF-β1-stimulated bioactivity was attenuated by changing IPF lung-relevant gene expression.[123] Referring to malignancies, it was recently reported that PFD induced apoptotic cell death in lung CAFs, and it increased apoptosis and synergistic cell death when combined with cisplatin. Takai et al. also reported that PFD inhibited cell viability and collagen production of CAFs, as well as the tumor growth induced by CAFs in breast cancer.[124] PFD could inhibit the EMT in NSCLC cell lines through inhibiting TGF-β signaling and TGF-β production. This concept is quite in contrast with using antifibrinolytic agents such as tranexamic acid, suggested since the 1980s, to suppress cancer cell growth, studied by Kikuchi et al., and in concert with my opinion on the same subject.[125] The mechanism of action related to these pro-clotting drugs seems to be inhibition of EGFR, and the mechanism of action for fibrinolytic agents' inhibition of TGF and FGF. In practice, I use these drugs depending on the findings in each case.

How does a tumor spread?

CTC

The entire concept of micrometastasis depends on the presence of these cells. These cells reflect the ability of cancer cells to *evolve*. The nature of these cells are not only different than the tumor cells, but also their stemness is identifiable based on the presence of the c-Myc marker. By measuring the PCR c-Myc, these cells can be identified and their activity correlated with stemness of the tumor cells. These cells mostly represent the progenitor cells, depending on the quantitative measurement of c-Myc. As such these cells are prognostic. The theory of bivalency from Stephen Baylin explains this section. The EMT transition plays a major role in the presence and maintenance of these cells (this is exactly why epigenetic therapies are supposed to be used as maintenance after remission; as such, using erlotonib is not indicated for stage I post resection lung cancer, except that it inhibits the EMT, at low dose).

Another target for EMT is SERPINA3, which induces EMT. EMT is also the main resistance pathway for immune checkpoint inhibitors (as well as AXL, TREM-2 and CCR-7), and as such AXL inhibitors have been suggested for use in ovarian CA along with both chemotherapy and immune therapies, with a PFS of 3 months!

Several studies on S500 B were aborted due to the fact that the cytotoxic therapies increased the number of CTC in patients with breast CA. That said, in a recent study by Francois in France, the CTC was suggested to be the indicator to decide which patients will benefit from hormonal therapies, compared to chemotherapy. This is contrary to the available data that clearly had suggested that the presence of CTC does *not* show a benefit from using chemotherapy. In the settings of adjuvant chemotherapy, certainly with the presence of the bulk of the tumor this is even more problematic than adjuvant setting. In the S500 B trial, the change of agents used from the first to the second line of chemotherapy in fact did not impact the negative outcome, and the CTC continued to increase despite the use of second-line chemotherapy, so as a result the study was prematurely terminated. Francoise also failed to show that PFS is impacted by CTC at all, which clearly disputes his argument in this trial.

Recent studies have shown the CTC identification based on EpCAM is inferior to the identification through other relevant markers, such as hexokinase in lung cancer. This marker has shown significant correlation in EGFR+ lung adenocarcinomas.[126]

The measurement of the CTC in early stage cancer can lead to the need for performing a whole body PET/CT staging if positive. For example, in the case of localized melanomas, the positive locoregional LN, or sentinel LN, points to the need for the staging CTs. Here we do not suggest the LN biopsy; rather, performing the CTC can give us the same indication, if positive.

This process is started with activation of NOTCH1 downstream to FGF-2 Wnt and Snail and Slug. The targets Snail and Slug are the most superficial targets at the tumor surface and they directly interact with the microenvironment. This crosstalk is extremely important to enhance the tumor's ability to be adopted by the host and continue the migration of the tumor. Tumor migration and invasion are necessary steps in metastasis. Targets for MMP are very important in this process. Inhibition of MMP is essential in inhibiting the tumor from migrating and spreading. By inhibition of EMT through nine different therapies (published by myself), you can completely stop the tumor's ability to spread. This step, along with inhibition of VEGF by epigenetic therapies, is extremely important in inhibition of metastasis. Studies show that median overall survival for patients with negative CTC in metastatic breast cancer is 47 months, compared with 29 months in CTC+ patients.

A new form of CTC has been described in breast CA, called regulatory brain cytoplasmic 200 RNA (BC200RNA). Also, there are small interference RNAs used in clinical applications to treat cancer, through peptide carriers.

One main concern about doing neoadjuvant chemotherapy versus in an adjuvant setting is the fact the since the bulk of the tumor is present, the risks of CTC dissemination is higher. As such, we see about 2 percent of patients with local breast cancer who end up with metastatic disease *while* they receive neoadjuvant chemotherapy.

ct DNA

The concept of genometastasis, which relates to the dissemination of tumor DNA into distant organs through hematological spread, is one of the major focuses of current research. Up to this point we were not sure how DNA shedding from tumor cells was able to spread to the blood, and if possibly they could infiltrate normal cells as a virus does and incorporate DNA into normal cell DNA. Now we know that through epigenetic aberrancies involving toll ligand receptors seven and nine (TLR-7, 9) the plasticity of the stem cells can change, and it also impacts the DNA's ability to enter into normal cells' genetic apparatus from tumor cells and make new colonies. cDNA can detect MRD in colon CA after therapy and can identify patients who may benefit from RNA vaccines in this group (Signatera Lab). There are toll-like receptors 7/8 inhibitors now in the trials combined with immune therapies in HCC. (One of the mechanisms of resistance to immune therapies is the TLR due to creating a cold microenvironment.) The MRD is suspected when there is suboptimal resection of LNs (less than 12) during surgery.

cDNA is generally removed by the blood nucleases and cleared by the liver, with a half life of less than 2 hours.

In general about one third of patients carry an actionable target in their cDNA. The presence of cDNA post surgery in breast cancer is as equivalent for complete remission as the MRI (75–85 percent specificity).[127]

New studies also suggest that the cDNA of HPV can be assessed as a marker for treatment response in head and neck cancer. The use of cDNA has also been suggested after neoadjuvant chemotherapy in patients who have bladder cancer with penetration to the bladder underlying muscle wall. Some patients may not need the bladder removal if the course of neoadjuvant chemotherapy has caused a complete response, with disappearance of cDNA.

The concept of tumor heterogeneity is dependent on heterogenous genetic markup and is readily measurable by identification of cDNA in the blood. Cancers such as basal cell carcinoma, melanoma, lung cancer, and bladder cancer have the highest heterogeneity.

As such, treatments that cause selective pressure and related advantages to the subpopulation of the cancer colonies should be selected wisely. For example, in breast cancer, the most common cancer in the world, inhibition of estrogen-driven breast tumors by an aromatase inhibitor or a SERD can induce triple negative breast cancer (SERD examples are fulvestrant, amcenestrant, or oral elacestrant or D-0502), or using Lupron or relugolix (an oral form of gonadotropin-releasing hormone) would give selective advantage only to the non-AR+ tumor cells.

Beyond the selective advantages that hormonal blockers produce for the non-ER cells, they also cause mutations in RB1 and further activation of PIK3CA and TP53. This is very novel information, as the old school of thought was that hormonal blockers could be safely prescribed to patients with ER+ breast CA. Now we know that addition of HDAC inhibitors reverses the transcriptional upregulation of c-Myc and Bcl-2 caused by hormonal blockers (tamoxifen) and restores endocrine sensitivity.[128]

Another example of selective advantage is tumor resistance after using chemotherapy, by selective advantage of TP53 subclones. This is why after each therapy in CLL, we test the FISH panel to discover the clonal evolution for TP53 and chromosomal abnormalities. (Deletion 17p, making them resistant to chemotherapies. Both Tp53 mutation and deletion 17p are classified as TP53 aberrancies.)

Further CDK inhibitors are used widely in breast CA, without looking at Ki-67, where studies show that their effectiveness is mainly beneficial in patients who have high Ki-67 tumors. Patients with high Ki-67 have a higher risk of recurrence (15 percent compared to 10 percent in breast CA). Additionally, oral taxanes (tesetaxel and Liporaxel, approved for breast and gastric cancer, respectively, in the US and Korea) have been used with P-gp inhibitors (encequidar) to improve absorption. These agents void the side effects such as neuropathies. A combination of taxane-based therapies with 5-FU seems the most effective chemo regimen in patients with gastric CA.

In prostate cancer, administration of estrogen alpha agonist has decreased PSA by 50 percent.

Also, detection of cDNA can be prognostic. Testing for ctDNA levels may help identify patients with colon cancer after surgery who will benefit, and those who will not benefit, from receiving chemotherapy.[129]

The application of liquid biopsies in general for GBM is unique. The main concerns about GBM stem from the fact that different GBM phenotypes—classical (with EGFR amplification), mesenchymal (with NF1 and

TP53 mutations), and proneural (with IDH-1 mutations)—all have different epigenetic makeups. For example, hypomethylation is commonly seen in mesenchymal form. That said, currently the only predictive factor in response is defined as MGMT methylation status (positive means poor response to temozolomide). Other factors are to be added to the prognosis, in my view. For example, the rearrangements of chromosomes 7 and 10, or TERT status, are prognostic, and more importantly the epigenetic code can define the response to the therapy and associated resistance. Liquid biopsies have provided the scientific merit to be used in GBM for diagnosis, prognosis, and response to the therapy. For example, in diagnosis, looking at different miRNAs can differentiate the metastatic brain lesions from the primary GBM. MiRNAs 10 and 21 are seen in GBM and mi-RNA 200 is seen in metastatic disease. Mi-21, a short noncoding RNA and a very important marker, is located at chromosome 7 and has been detected in several research studies by different technologies (TaqMan, TargetScan, etc.). The importance of this miRNA relies on its oncogenic activity to impact VEGF, PTEN, and apoptosis (studied in lung, breast, and colon CAs). It is postulated that miRNA 21 can be targeted by both a synthetic antibody and curcumin.[130] There are ongoing studies on formulating the nanoparticles of such therapies delivered to the brain.

Long non-coding RNAs are used diagnostically in research and correlated with prediction of breast cancer recurrence by MRI radiomics.

Several liquid biopsy techniques can be used to identify the presence of tumor signature in the blood (plasma). One is CTC. These cells do not exhibit the standard EpCAM markers, as other such signatures need to be used. For example, using CTC1 technology looking at antibodies (anti CD45, 14, and 16) detects the molecular signature for mesenchymal cells (SERPINE1, TGFB1, TGFBR2, and VIM).

In my experience, the detection of CTC has been possible in some cases of gliomas, by looking at c-Myc. My experience is not good in the detection of cDNA through PCR. As such, I can not recommend one way or another on the predictability of the cDNA for detection. That said, I believe there is significant focus on detection of any signature for histone 3K27M, as it correlates with high-grade GBM.

Another form of liquid biopsy is tumor educated platelets, which can carry these miRNAs to the blood, and further can be used for diagnosis. The presence of such markers, including CTC, can differentiate between real progressive disease and pseudo-progression.

Furthermore, urine and CSF can be used for detection of both cDNA and microRNAs. Other markers can include 2HG (hydroxyglutarate) for IDH-1+ GBM in urine.

Commercially available tests are offered by many laboratories: Guardant360, Genomics Lab, Foundation Medicine, Elevation Oncology, and more. The FoundationOne assay has been approved as a companion diagnostic for Piqray in breast cancer, rucaparib in ovarian cancer, and alectinib in non-small-cell lung cancer.

Based on the theory of bivalency presented by the Epigenetics Dream Team's Dr. Stephen Baylin at John Hopkins, circulating tumor cells have the ability to differentiate into normal cells or become more plastic and become cancer stem cells depending on two key genetic transcription, E-cadherin and c-Myc. c-Myc represents the stemness and E-cadherin represents the epithelial, more differentiated daughter cells. Thankfully, with the new technology that is available we can identify the mRNA expression of the circulating tumor cells. As far as their c-Myc transcription and such, we can know how much these circulating tumor cells are stemmed. You can also identify their ability to divide through telemeters and CK transcription. The clinical management by a traditional oncologist is defined based on shrinkage of the tumor as a sign of response seen in images and measured by RECIST criteria.

CTCs also can be used to assess the mRNA expression of different genes; for example, they can show AR-V7 in patients with metastatic castration-resistant tumors. ARV targets (TMPRSS22, KLK2, KLK3, and FOLH1) as well as neuroendocrine targets can be assessed in prostate cancer CTCs (Journal of Clinical Oncology, Sept. 2021).[131] I personally assisted a laboratory in Germany called BioFocus in establishing a testing platform for both DNMT and HDAC epigenetic markers in CTC. Since my collaboration with BioFocus, this test has become available and is widely used by cancer centers around the world.

If a patient responds to specific treatment and the size of the tumor decreases this is called progression-free survival, meaning that for a certain period of time the images do not show increased size; rather, the size can decrease. This mostly represents the activity of the daughter cells or, better said, the number of daughter cells. The PET scan, however, measures the activity of these cells and is a better criterion for metabolic activity than size. Overall survival, however, has nothing to do with the size of a tumor, meaning that it's not related to the number or the activity of daughter cells. Overall survival is an indication of the activity of the stem cells, their ability to produce new daughter cells, and the aggressiveness of these cells, which is again measurable through existing biomarkers. Metabolic activity of higher than 7 is identified as significantly elevated (for example, in CLL it could be a sign of Richter's transformation).

The concept of the Warburg effect, describing the anaerobic glycolysis, relates to the conversion of pyruvate to lactate, in excess of tumor hypoxia, and the activation of HIF where it leads to tumor fermentation, increased LDH and tumor access to synthesis of lipids and nucleic acids necessary to maintain the function of stem cells. Tumor activation of hexokinase and metabolic access to glucose happen in parallel to activation of PIK3CA and Akt/EGFR. As studies by Eric Raabe MD, PhD at John Hopkins have shown, using a modified version of DON (6 diazo-5-oxo-L-norleucine), JHU395 has shown effective glutamine metabolism disruption, inhibition of the fermentation process, and Myc driven neuroblastoma cells in pediatric medulloblastomas.[132] I am concerned about new approval for Lu-DTPA in these patients and its effect on this gene. Prior studies by Thomas Seyfried (published in his book *Cancer as a Metabolic Disease* and recently reviewed in *Oncology Times* in 2021),[133] referred to cases treated with ketogenic diet, specifically glioblastoma inheriting IDH-1 mutations, with positive response. It is in my opinion impossible to generalize this phenomenon of response without an epigenetic switch to permit the mitochondrial use of glutamine, and this has been validated by studies using microRNAs.

The role of epigenetics in giving the green light to this type of cascade is important as it resolves the entire argument between scientists who defend the concept of a tumor as a metabolic disease and scientists who believe that a tumor is a genetic disorder. In other words, we can now say confidently that cancer is an epigenetic disease, as it switches on the metabolic pathway leading to cancer growth in the presence of genetic mutations or absence of such a predisposition. Genetic mutations only represent a co-finding in this scenario, and resemble accelerated growth only if the epigenetics are aberrant.

There are so much effort conducted on metabolic treatment of cancer, from using Bcl-2 inhibitors and identifying NAMPT in TNBC, using inhibitors of NAMPT to hexokinase inhibitors, and 2-deoxyglucose, and dichloroacetic acid. Many of these therapies are effective only if used with epigenetic therapies. For example, LDH phosphorylation is inhibited by DCA, but unless the ubiquitination site is also inhibited, the LDH does not respond to DCA alone.

These prognostic surrogate markers of survival all have to do with the activity of the stem cells. For example, growth factors are prognostic, or markers for angiogenesis are prognostic, meaning that they correlate the patient's overall survival. In the setting of colon CA, Heinz- Josef Lenz et al. showed that no prognostic/predictive association is found for any signaling pathway (except the angiogenic pathway!), as such only few biomarkers of angiogenesis were identified in CRC.[134] These markers are ANG-2, CA-9, CEA cell adhesion molecule (CEACAM1), collagen-IV, IGFBP-1, intracellular adhesion molecule (ICAM-1) and u (urokinase receptor) PAR, placenta growth factor, and platelet-derived growth factors. Drugs such as Seralutinib that are used in pulmonary hypertension can be considered for off-label use if PDGF is increased.

Another example, microRNA 1275, is related to increased IGFBP2 and IGF-1 metabolites, and it can be reduced by taking EGCG, quercetin, and Tamarix. Methyoxylated quercetin also reduces microRNA 15 and 16 related to p16 and TP53 mutations.

Recent medications are targeting these growth factors; for example, Nintedanib inhibits PDGF/VEGFR and FGF2. Pemigatinib is approved for FGFR+ cholangiocarcinomas, and Bemarituzumab, as monoclonal antibody; in gastric cancer, harboring FGFR2b, with only 2 months improved PFS. Ingigratinib also works on the same target, but a small molecule. Another medication for gastric cancer is BOLD-100, a ruthenium-based small molecule that alters unfolded protein response.

Recently there have been efforts in producing drugs that block the growth factors, such as FGF2, TAS 120, or Futibatinib, in metastatic breast CA and others.

Other new biomarkers include mesothelioma soluble related peptide (for both mesotheliomas and ovarian CA), and IL-2 soluble receptor (for lymphomas) and as such there are drugs in the trials to inhibit this prognostic and important target. I always check Il-2 soluble receptor in lymphoma patients with a high predictability of response to any therapy. It is also important to know that if elevated, this marker could relate to predicted failure to chemotherapies.

But these markers all are also influenced through epigenetics and they go parallel with stem cell activity. The more a tumor is active the more it is able to spread and the more it can penetrate into the blood vessels.

There are certain mutations that dictate the epigenetic regulation of proliferation; for example, mutations at CTCF binding sites have been associated with multiple forms of cancer, disrupting the epigenetic regulation of the disease.

A process called EMT or epidermal mesenchymal transition again is under the control of stem cells and the epigenetic influence that controls the stem cell activity pretty much can also control the angiogenesis and the ability of the cells to spread. There are also specific enzymes that are regulated through epigenetic influence and these enzymes are in charge of the metabolism of the cancer cell. For example, the metabolism of glutamine is completely dependent on epigenetic influence; also, the signals translate into the growth of the tumor and crosstalk between the tumor and the host microenvironment is under epigenetic influence.

It is important to understand that since it is under the control of epigenetic influence we can modify and reverse the process by using specific treatments that target the epigenetic code. This subject is a mainstay of many treatments in the epigenetic field.

Recently scientists have studied cancer genome studies and there has been discovery of a four-drug combination in suppressing a protein called Raf kinase inhibitory protein, without activation of feedback loops.[135] These medications are JNK inhibitor SP600125, MEK inhibitor Trametinib (GSK1120212), MLK inhibitor URMC-099, and CK2 inhibitor CX-4945 (Silmitasertib).

Finally I would like to emphasize the role of cDNA in identification of not only somatic DNA mutations, but also copy number variations and more importantly the epigenetic changes recognizing the tissue of origin (90 percent) with minimal false positivity (0.5 percent), through liquid biopsy. The sensitivity is less as the stage is less, 16 percent in stage I.

As such, by performing screening tests through liquid biopsy, it is

	Mutation	Copy Number Alterations	Methylation
Cancer Detection	✓	✗	✓
Low False Positive Rate	✗	✓	✓
Cancer Location	✗	✗	✓

speculated that the 5-year cancer mortality will drop by 20 percent, as there would be a 50 percent reduction in late stage cancer diagnosis.[136] These claims are presented by Galleri for the Grail test; as I mentioned earlier in this book, I am *not* a fan of this test. In my review the circulating microRNAs are much more predictable for assessment of variety of solid tumors both for prognostic as well as treatment response indications. Currently there are studies looking at c miRNAs in germ cell tumors with very promising results, to the point that it is suggested that soon it will replace the standard CT scanning serial monitoring for surveillance, in clinical practice.

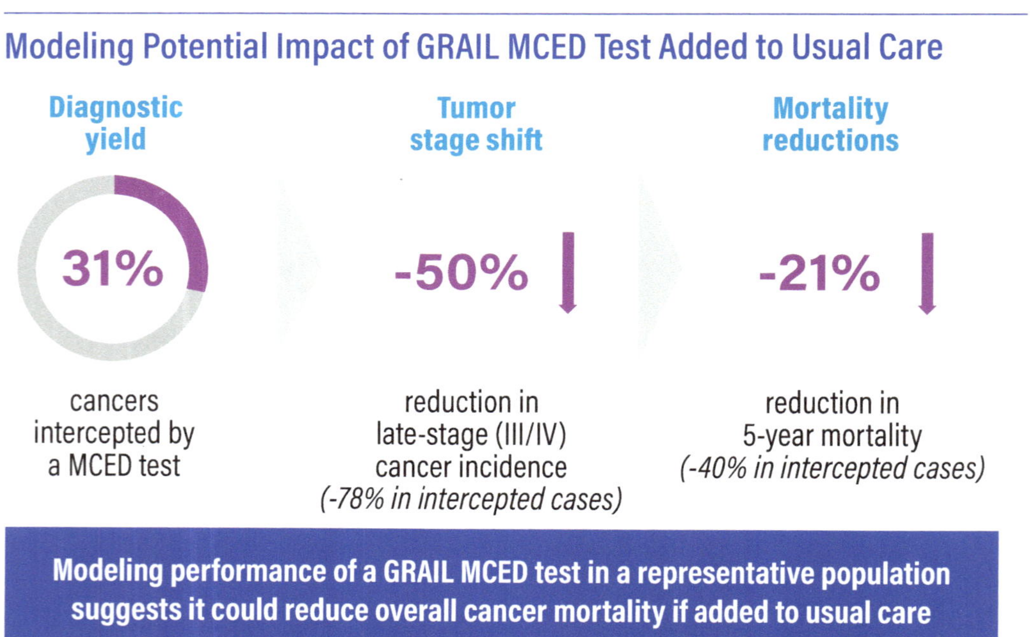

Chapter Ten

Active Surveillance/Watch and Wait, The Rule of First, Do No Harm

I THINK NOW IT IS TIME TO DISCUSS WHY DOING NOTHING IS AT times the best thing an oncologist can do! I would only say an oncologist because I think we can all agree by now that using many conventional methods of care can only exacerbate the circumstances by increasing the cancer stem cell activity. As we discussed earlier, heroic brain hemispherectomies for gliomas, or the Halsted method of mastectomy for treatment of breast cancer, are now outdated and we have realized that cancer is not a local disease, but rather a systemic problem. New data even suggests that axillary lymph node dissection after mastectomy in patients with positive axillary sentinel nodes does *not* increase patient survival and in fact increases patient morbidity, with no benefit. One study on the impact of axillary dissection among patients with positive sentinel node by James Sun et al. said that new magnetic compounds are used to identify these nodes;[137] as such, the concept of sentinel node biopsy is questionable in my mind and it not recommended as a standard procedure in my practice. Also, adjuvant radiation is never recommended routinely, as the MA39 trial is currently evaluating (questioning) the radiation benefit depending on the genetics of the tumor with limited node involvement.

In melanomas, at stages I and II, surgical resection is the only recommended therapy. New studies show that wide excision does *not* improve patients' survival when the disease is at lower risk (when there is lack of ulceration or LV invasion) and sentinel LN biopsy is only recommended if WE is suggested (a tumor more than 2 cm T2-4). The same concept is true for head and neck cancers, where neck dissection seems to be beneficial only for tumors over 3–4 cm in size. In some studies, neck dissection of lymph nodes has no survival advantage over sentinel node biopsy.[138]

As we move forward with the application of knowledge in our practice, we have stayed away from whole-brain radiation (in my practice) and even nontargeted radiations, intraarterial chemotherapies (which show no evidence of improved survival in different kinds of cancer, such as liver CA), and further unnecessary interventions. Other conflicting studies include a systemic review published by ASCO in 2020, which denied a benefit from chemo drugs in a neoadjuvant setting of breast CA at stage I.

In a wide range of tumors where growth factors are overexpressed the application of radiation causes EGFR overexpression, which translocates to the nucleus where it acts as a transcriptional factor. It is also suggested that radiation application in conjunction with consumption of arginine can significantly enhance its efficacy, due to the overstimulation of NO synthases (Oncology Times, Dec 2021).[139] As such, we recommend intravenous arginine for patients who desire to receive radiation specifically for the metastatic CNS disease.

The watch-and-wait policy is known for prostate cancer when the PSA doubling time is more than 24 months. The same concept applies for rectal cancer when a patient is treated with neoadjuvant chemotherapy. This knowledge has helped us significantly in avoiding harmful surgeries in this population.[140] Also, the application of neoadjuvant radiation and then surgery has been a question as well, as the time interval of a few (6–8) weeks been proposed, but as we discussed tumor regression has failed to be an independent prognostic factor and a poor proxy for long-term survival!

We know now that even with the most radical surgeries the disease still remains at a microscopic level and the addition of surgery can only *debulk* the tumor (if it persists after chemotherapy), and will not impact the overall survival of the patient, even if the tumor PFS is improved. In lung cancer, for example, surgery has been long claimed to be indicated in all stages, but we know that in stage III, the surgery has shown minimal results, with great risk of residual disease. We also know that many tumors at lower stages have micrometastatic disease and already are at stage IV, which makes surgery futile in improving patients' survival. This is why the data on recurrence post surgery has such a wide range (8–60 percent in lung cancer) and using adjuvant chemotherapy only adds an improved survival rate of 10 percent![141] (Clinical Advances in Hematology & Oncology, Oct 2021)

This is a great example of the discordance between PFS and OS. The sustainability of a complete response obviously is measured by the microscopic presence of the tumor cells in the lymph nodes measured in subsequent scans. If a patient maintains a nondetectable presence of tumor recurrence in 1 year, the sustainability of CR is about 88 percent. After 3 years it is about 97 percent, and after 5 years about 98 percent. This model gives us a great application and transformation of knowledge to other types of solid tumors, such as breast cancer, where the question of duration of maintenance therapies is challenged.

For example, in my clinic, we advise a low dose of tamoxifen (5–10 mg), which patients can choose to take for 5 years, but the data now reveals that most recurrences happen in the first 12 months. Tamoxifen has been studied in the prevention of cancer, at doses such as 2.5 mg, for reducing the density of mammograms. Similarly, some studies suggest that using estrogen in postmenopausal women reduces the chance of breast cancer and its related mortality (Breast Cancer Prevention, Time for Change, JCO Oncology Practice Dec, 2021).[142] In some studies tamoxifen has been proposed at once a week dosing for the same purpose. My rationale for using tamoxifen is not ER receptor inhibition, but rather its effect on telomerase. Also, the application of adjuvant chemotherapy for patients with rectal CA (with Xeloda plus Ox, for stage III, or even Xeloda alone (four cycles for stages I and II), has shown to be as effective as the FOLFOX regimen for six cycles), or can be used for 12 months.

The same concept as for epigenetic therapies applies when used for preventive measures (secondary prevention), with the further organ sparing approach safely used for stages I and II rectal CA. That said, keep in mind that local recurrence may be higher in this setting, compared to standard therapies, but no difference in OS is observed.[143] We also know that CT findings in these patients have shown no correlation with survival, but the PET finding in the primary tumor (not metastatic) has been prognostic (Femke CR Stall, et al.).[144]

Watchful waiting or active surveillance was the term used for monitoring prostate cancer when the disease is best managed by doing nothing! The reality is that the standard approach has caused so much worsening of the outcome, both in reducing quality of life (from radiation damage to ADT causing osteoporosis and cardiac disease) and in stimulation of stem cells and reducing survival, that at one point, watchful waiting would become the standard of care for a larger set of patients. Early on when this term was introduced, it was suggested for patients with a PSA doubling time of more than 24 months.

Now it is suggested for patients with a PSA doubling time of more than 9 months. In other words, more and more patients are candidates for observation alone. Also, studies show that for localized disease, with fewer than three LNs involved, watchful waiting is an appropriate choice, when in the past positive LNs meant active therapy. Additionally, therapies that initially were thought to be effective in combination, such as AAP, are now questioned for patients with lower Gleason scores (less than 8) and in fact many oncologists try to avoid abiraterone when the number of LNs is less than three.

Our strategy of watchful waiting or active surveillance (as different as it may be) is based on using metronomic therapies to reduce the residual disease. This approach has been investigated in both rectal CA and breast CA, as discussed. For example, in rectal cancer, the omission of surgery and radiation is suggested in selective cases. This is while the standard of care is still using neoadjuvant radiation and chemotherapy plus surgery plus adjuvant radiation, but many studies have shown that radiation only reduces the local recurrence rate and does not affect the patients' survival.

On using metronomic chemotherapy, for example, a well-organized study in Japan, published in *The Lancet Oncology* in January 2021,[145] showed that in patients with ER+ breast CA, using Xeloda or S-1 (tegafur or gimeracil) benefited the patients after accomplishing the complete response as a metronomic use, causing disease-free survival to be increased compared to the control (hormonal blockade alone). This approach has also shown good results in head and neck cancer after definitive radiation and chemotherapy.

The concept of *cytotoxicity* and the expectation that the more cytotoxic a drug is, the better it works on cancer, is completely disputed. Even in the setting of "residual disease" after chemo and radiation or surgery, Capecitabine has shown superior results to more cytotoxic chemotherapies like platinum in triple negative breast cancer.

The metronomic schedule can be a 14/21 schedule. There are similar suggestions for using metronomic Ibrance in patients who have accomplished CR, as well. This is a clinical shift from using these drugs in later settings versus as a first-line therapy. It has been shown that patients who were treated with chemotherapy before CDK inhibitors had poorer outcomes than the ones who did not receive chemotherapy. In this setting (long-term use) Verzenio is preferred as it has a reduced risk of myelosuppression compared to Ibrance and Ribociclib/Kisqali. This is why this medication is the only CDK inhibitor approved for early breast cancer along with AI (or tamoxifen) as an adjuvant therapy when the Ki-67 is high. When this data was published and the drug received approval, the first thing that came to my mind was the fact that the use of the drug can be considered in other types of tumors where the Ki-67 is high, and consistent with the data showing that CDK inhibitors can be used along with Temodar in gliomas. The data is yet to be out for dalpiciclib, but I am not optimistic.

This was the second study, in my recollection (the first study was using letrozole plus oral Cytoxan), that confirmed that the combination of chemotherapy and hormonal blockade is not only feasible but also improves survival in patients with CR in breast cancer. As we know, oral 5-FU or Xeloda has been used widely in a metastatic setting or in TNBC, but use in an adjuvant setting in stages I through III was explored in this study, with a good outcome (4 percent improved response rate).

There are also studies on using Xeloda along with oral Taxol (tesetaxel) in both an adjuvant and a metastatic setting. We recommend performing DPYD*2A genotyping in clinical practice before starting patients on Capecitabine. We also know that in defective mismatch repair gene colon CA (early stage), using 5-FU harms the patient.

Other oral combinations of chemotherapeutic agents are Temodar and topotecan, which seem to be synergistic in certain types of cancer, including sarcomas.

Chapter Eleven

Tumor Localized Therapies

As we know, the tumor and the cancer are two totally different things. As I see it, the tumor is only the symptom of the cancer. The majority of conventional efforts have been directed on treatment of the tumor and not of the cancer itself, meaning that what caused the cancer from beginning, the stem cells, are untouched by conventional care. This is why the cancer presents itself with recurrent disease (tumors) after each therapy. That said, in this section I would like to discuss some localized therapies used to treat different types of tumors, with marginal efficacy.

1. Intralesional treatments with BCG. A live attenuated strain of BCG is injected into the lesion, with 19–45 percent pathological response/complete response causing tumor necrosis and T cell infiltration. This concept is very similar to injection of BCG into the vesical bladder in early onset bladder carcinomas. This also can be done into the LNs, although the data on efficacy is not clear.

2. IL-2 injection inside the lesions. This can be done as a form of inhaled CSF in metastatic lung cancer, which has been studied in RCC and sarcomas, or in melanoma in the form of intratumoral injection, with an overall response rate of 15 percent. It is costly and can have inflammatory side effects, at times very severe. (There are recent suggestions on using systemic drugs such as CSF inhibitors systemically in sarcomas.)

3. Rose bengal is a water-soluble xanthene dye used in diagnostics in eye care. It is pretty much a similar mechanism to a PD-L1 inhibitor and has in fact been used with a combination of immune therapies, with different ranges of response.

4. Talimogene laherparepvec is a genetically modified herpes simplex virus type 1 used as an oncolytic therapy to treat locally advanced melanomas, and in a different version is studied in gliomas.

5. Isolated limb perfusion and infusion. Melphalan and TNF have been used as very highly toxic therapies that are heated and delivered to the limb to avoid systemic toxicities. They are used in a similar fashion to the intraarterial infusion of chemotherapies used in liver cancer and other types of cancers, with limited efficacy.

6. Carbon dioxide laser ablation. Imiquimod and topical diphencyprone have also been used as localized immune therapies. These therapies have shown significant results in lentigo melanoma. The idea is to avoid amputation, which we unfortunately see very commonly in cancer patients with melanomas.

7. Photodynamic therapy. Padeliporfin (an immune photo-activated cancer therapy) offers IV delivery of an inactive drug, which when activated destroys the blood vessels around the tumor. It has been approved for urothelial carcinomas.

8. HIFU for prostate cancer, with a good 70 percent on average 2-year PFS (Prostate Cancer, review article Mitch Hayes).[146]

9. Irreversible electroporation, or generating pores in the cells by electrical pulses. This technology has a better side effect profile than robotic assisted prostatectomy, with less urinary incontinence.

10. Focal radiofrequency, ultrasound, and laser ablation.

Chapter Twelve

Financial Aspects (Toxicity) of Cancer Therapies

THE COST OF CARE FOR CONVENTIONAL THERAPIES IS ENORMOUS. This trend is increasing and in fact, new drugs are increasingly more expensive. Compared to 2009, we have doubled the sales revenue from cancer drugs, and which exceeded USD 103 billion in 2019 alone (Congressional Budget Office, Research and Development in Pharmaceuticals, April 2021).[147] More than 119 new drugs have been approved for cancer, accounting for 29 percent of drug approvals.

About one third of patients with cancer report having financial burdens. The ultimate financial "toxicity" of the therapies affect the patient and the community as a whole. When financial incentives are not aimed at quality of care, the services are only incentivized by quantity of care, and that is simply the number of patients treated and number of drugs prescribed. Instead of using the minimum amount of therapies with similar results, the incentives are on marketing combination drugs and maximizing sales. The most obvious amongst them all is the combination of the most expensive protocols, of antiangiogenic drugs such as Avastin with PD-1 inhibitors, which is suggested for wide ranges of tumor types, such as hepatocellular CAs and colon CA (or levanitinib plus PD-1 inhibitor for endometrial CA).

The same concepts prohibits from effective preventive measures, where spending could be tailored by preventing the disease and improving and promoting health through a simple measure such as taking ASA and vitamin D, which has shown a prophylactic role in preventing 85 percent of colorectal cancer. Other studies have suggested that one etiology for lymphomas is poor exposure to sunlight before the age of 2, and the addition of vitamin D can improve patients' response in lymphomas. Yet it is still not recommended by ACS because of "unknown risks." The risks of taking vitamin D are not known? What risks are there that are overweighing the benefit of reducing cancer risk by 80 percent?

We already know that 45 percent of cancer cases are caused by modifiable risks such as diet, smoking, and obesity. In this regard, only 5 percent of NIH's budget annually is spent on prevention, and indeed we have made almost no progress in detecting cancer early, or preventing it by using effective screening tools. In California only 0.9 percent of people at risk of lung cancer are being screened! Now, if you would like to laugh, you can also look at a study published in Hem Onc in August 2021,[148] where the article suggests that obesity "improves" the survival rate in prostate cancer. I will just leave it there. I think the author (Alverto Martini, MD!!) has forgotten that cancer incidence increases with obesity, so the discussion of survival would be irrelevant, and how could someone in their right mind encourage a patient with cancer to get "obese" to have a better chance of survival! This is a perfect example of pseudoscience, in my view.

Simple solutions, such as the concept of early detection by using breath analysis, was studied by University of Latvia, the same concept that dogs can smell the cancer through breathing in of the affected individual (25 volatile organic compounds, mainly toluene, xylene, benzene and sulfur). The technology is patented under ASdevice and eTrace medical technology. These electronic "noses" use DNA nanotube peptides to detect different kinds of cancers, with very good predictability (90 percent for pancreatic and ovarian CA) (Erica Carpenter, Liquid Biopsy Laboratory).[149] Other detection technologies such as fragments of DNA (called DELFI) are under investigation.

The cost of maintenance therapy with an EGFR inhibitor such as osimertinib is about USD 200,000 every year and although there is no evidence that it can improve patients' survival if used in the early stages, it is currently recommended to be used at all stages in non-small-cell lung cancer. Some of these patients never recur and yet are treated for decades with this high-dollar ticket. That said, if a therapy is deemed effective, the financial cost should not be a barrier for the care. That said, osimertinib is recommended as part of neoadjuvant therapy for at least 3 years after platinum doublet in EGFR+ tumors. In patients with EGFR L858R mutations, however, it is suggested that using erlotinib and ramucirumab in combination is superior to osimertinib a in metastatic setting (especially in Asian patients). The cardiac toxicity of osimertinib is reported at 4.4 percent, so it remains a concern. In patients who have progressed after osimertinib, amivantamab and lazertinib have been suggested. Patritumab deruxtecan (an anti-HER3 IgG1 preloaded with TOP1 inhibitor) also has been offered. Finally, BLU-945 is a fourth-generation TKI.

The same is true about using expensive immune therapies for 1 year after chemo and radiation in unresectable stage III lung cancer. How can one prove the efficacy of immune therapies as maintenance treatment after chemo and radiation has caused complete remission (there is no control)? Furthermore, due to serious adverse effects (100 percent reported) in AML, their use is questionable. Immune therapies also showed no benefit in the IMagyn050 trial in addition to the standard of care for improving PFS.

There is also no improvement in combination with adjuvant therapies in colon CA, as there is no benefit from adjuvant chemotherapies when MSI is high. In stage III colon CA 50 percent of patients will do fine without chemotherapy anyway, and now the rationale is to use 3 months instead of 6 months of therapy and use Xeloda instead of 5-FU. The same is the case for stage II high risk, where 3 months of CAPOX is as equally beneficial as 6 months of FOLFOX. Combination therapies also have failed to provide meaningful benefits in many other settings, including the use of three targeted drugs at a time; for example, cetuximab has failed to improve survival when added to BRAF and MEK inhibitors, in the BEACON study. This obviously depends on what type of BRAF mutation is there; for example, BRAF V600E type is associated with CIMP and hypermethylation of the DNA promoter regions and gene silencing. RAS mutations are more frequently seen in non-V600E phenotype, and chemotherapy advantage is more pronounced. Fewer right-side and highly graded tumors are seen in *non*-V600E types as well. MSI phenotypes are more frequent in V600E types.

Biosimilars are nonbiological original medicine with an active ingredient that mirrors the reference biologically and has no clinically different values. So far there are few targeted drugs approved as biosimilars; Avastin, rituximab, and Herceptin are examples of this category of drugs. There are 73 candidates waiting for approval at the FDA. The reason is that the hospitals sell the drugs to the insurance companies at a very high price, while the co-payments of patients would pretty much be enough to cover the cost of biosimilars.

The major financial toxicity of cancer therapies comes with the risk of hospitalization, which is obviously greater in patients who are older, as well as those with comorbidities, low kidney function, or low albumin. In general there is at least one episode of unpredicted hospitalization in one fourth of patients with cancer. Overall, the disease burden is highest in the US than in any other country. Disease burden, which accounts for both premature death and years living with disability, is often measured using disability-adjusted life years. Though DALYs have declined in the US and comparable countries since 2000,[150] the US continues to have higher age-adjusted rates than those of peer countries. In 2017, the DALYs rate was 31 percent higher in the US than in other comparable countries, on average.

Amongst all other conditions, patients with cancer have the highest out-of-pocket costs. This can also include supplements used for cancer, an industry of about USD 40 billion annually. Some of these supplements are not studied well but unless they pose a real threat to the public they cannot be removed from the market.

The following diagrams summarize the difference between conventional oncology and the outside-the-box approach. Certainly, the common small areas shared between the four circles seen on page 163 represent the medicine available in clinics. So, 90 percent of translational science is not available to patients, as you can see, because either it is stuck in regulatory approvals, or the marketing is not strong, or insurance does not cover the cost. This is the standard of care. Sadly, it shows that the role of science is as important as marketing or other elements (circles).

In the second diagram, you see that translational science is the only part that stays the same, with just minimal changes. It basically resembles mathematical biology, so it is more specific in its predictive value of the genomic mathematical share. Otherwise, it pretty much replaces the other circles. The common sense of every physician trained in oncology defines what is harmful and what is beneficial. Most of the "approved" oncology products are more harmful than beneficial, so much so that regulatory agencies like the FDA pull their own "approved" products from the market every day. One of the best examples is Avastin, which was pulled out of the market

Average out-of-pocket spending per person based on diagnosis status, in U.S. Dollars, 2013

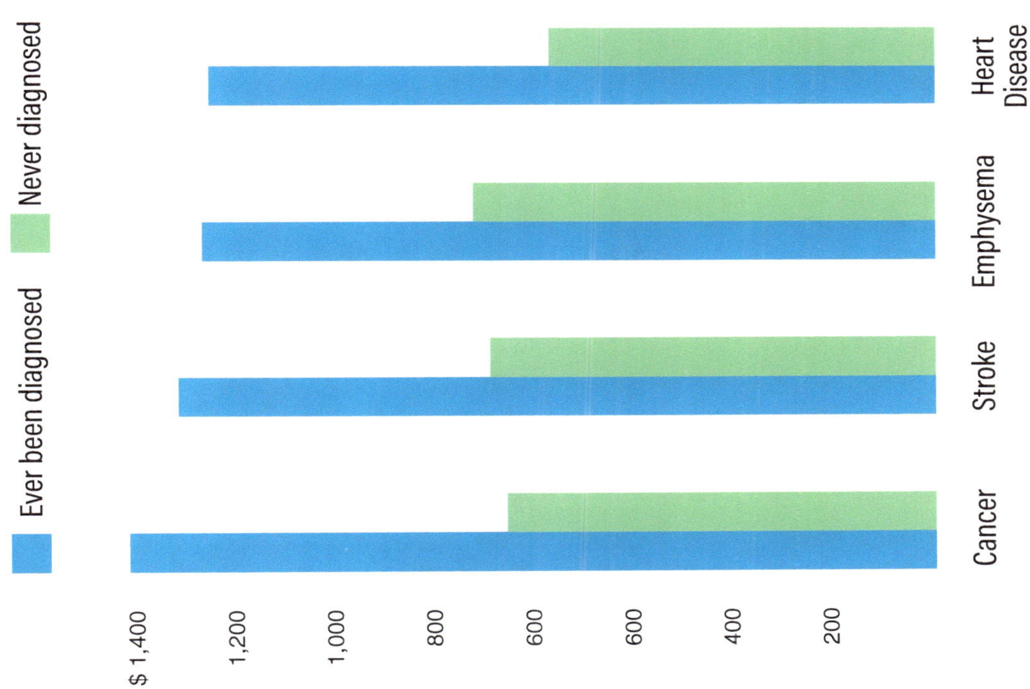

Disease burden is higher in the U.S. than in comparable countries

Age-standardized disability-adjusted life year (DALY) rate per 100,000 population, 2017

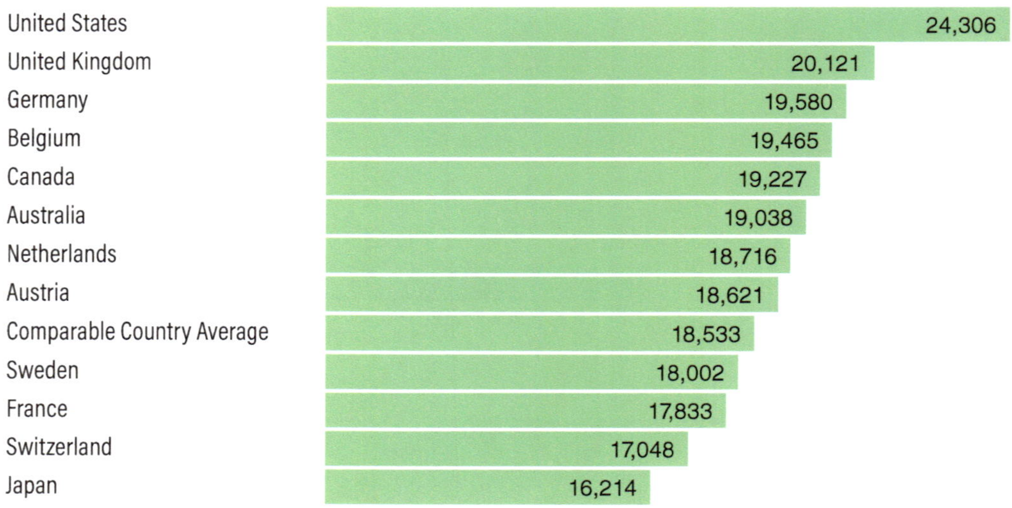

The U.S. has higher than average disease burden from lung Cancer

Age-standardized Disability-Adjusted Life Years (DALY) rate per 100,000 population for lung, tracheal, and bronchus cancers, 2016

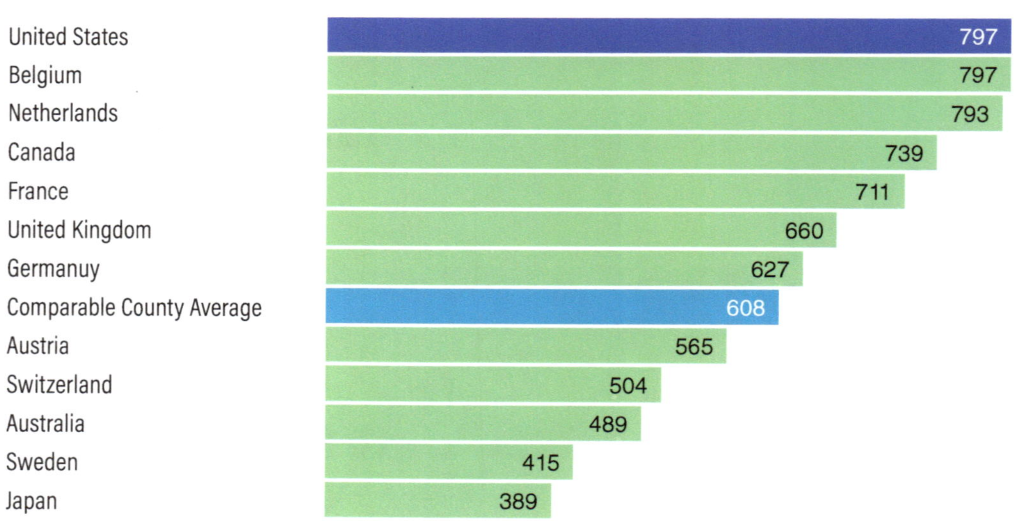

(disapproved), after being approved and used for a while in 2005, due to its harmful effects, causing thousands of patients to die, per the FDA's own admission.

This is exactly why public confidence in the NIH and the FDA has dropped significantly (it is now at 37 percent in 2021, compared to 42 percent in 2009). The same is true for the CDC, whose judgement and recommendations are trusted by only 50 percent of people.[151]

Other examples are discussed in this manual, including the fast-track designations or conditional approvals, where the condition is never met and the drug only can be pulled out voluntarily by the manufacturer (which never happens). As for marketing, I am opposed to marketing oncology products at all. In fact, using false advertising and giving patients false hope is extremely dangerous, but still a common practice in the US, to the point that the advertisements do not include transparent information. An ad will visually capture the audience by showing something like two people dancing at the beach, while it shows only a small disclaimer about side effects, including death, in small text at the bottom of the screen! The standards of marketing are below reasonable in the oncology world and everyone knows it. I would replace the current marketing with meaningful transparent public education where the audience has a chance to ask questions and inquire about their concerns. We have again discussed in detail in this book that only 1 out of 4 oncologists are transparent or factual about the medications they prescribe to patients, and 3 out of 4 do *not* accurately report the medications' toxicity to patients. Furthermore, when an oncologist has a diagnosis of cancer himself, only 1 out of 10 chooses to receive the medications they have prescribed to others! Patients need to be aware of these discrepancies and fraudulent claims.

As far as national impact on health care, it is obvious that the majority of financial toxicity related to the treatment of cancer is related to metastatic disease. For example, just in breast cancer, at stage IV, it is predicted that in 2030, there will be USD 152 billion per year for cost of care in the US alone, which is a 140 percent increase from today.

Finally, regarding the insurance coverage, I will comment on the most simple principle of medicine: *The outcome should define the incentives*. If a provider is incentivized by improved patient outcome, there is a strong reason to improve care and focus on results. On the other hand, if the incentives are not based on positive outcome (as is now the case with insurance companies), when a patient dies, the provider is paid the same as if the patient had been cured. This is why oncology care, per my principles, should never rely on income received from the patient; rather, it should be dependent on patient satisfaction and outcome. As such, since the purpose of the care is divine, it should only be based on a nonprofit structure (either completely or partially). This is completely opposite the current system, where a provider is paid even after the patient is referred to hospice care (an unethical kickback). A recent study in *The Oncologist* (Vol 26, Sept 2021)[152] suggested that certain practices of oncologists are influenced by financial incentives (the influence of the industry), which directly impacts their practice guidelines and NCCN membership. In other words, the more they are financially motivated, the more these practices are inclined to adopt incentivized protocols and guidelines.

Since the intention of this publication is not to replace the formal training required for so-called oncologists to be prepared to understand my therapy, I have only introduced concepts to trigger the interest of open-minded providers, who may then further expand their knowledge by studying my more than 100 publications and referring questions to me.

In the end, I hope the reader appreciates the fact that the intention of this manual is also to engage the reader in a reality check and further promote public education for patients to be their own advocates, and to further enhance the concepts and disclose the misconceptions in oncology care as it is practiced today.

Thank you.

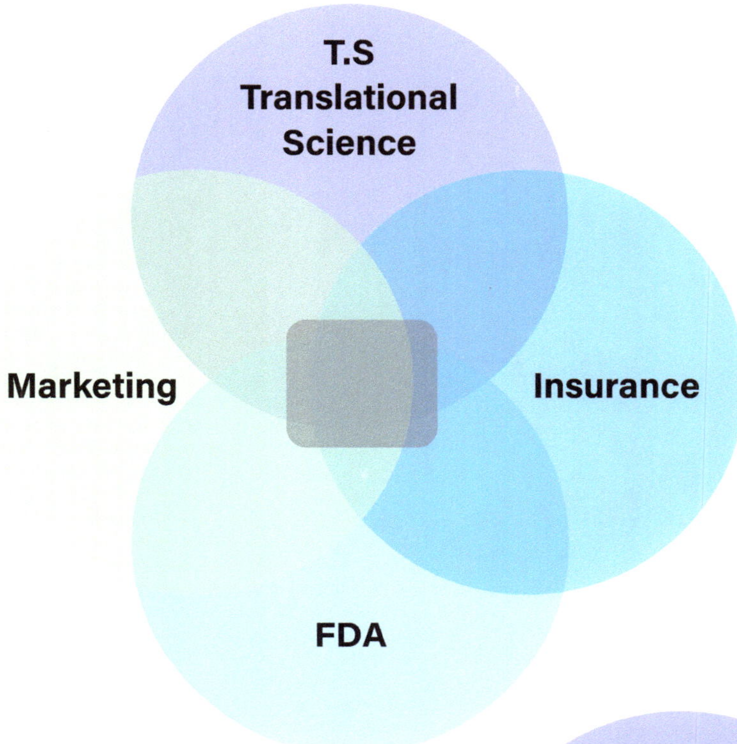
T.S
Translational
Science
Marketing
Insurance
FDA

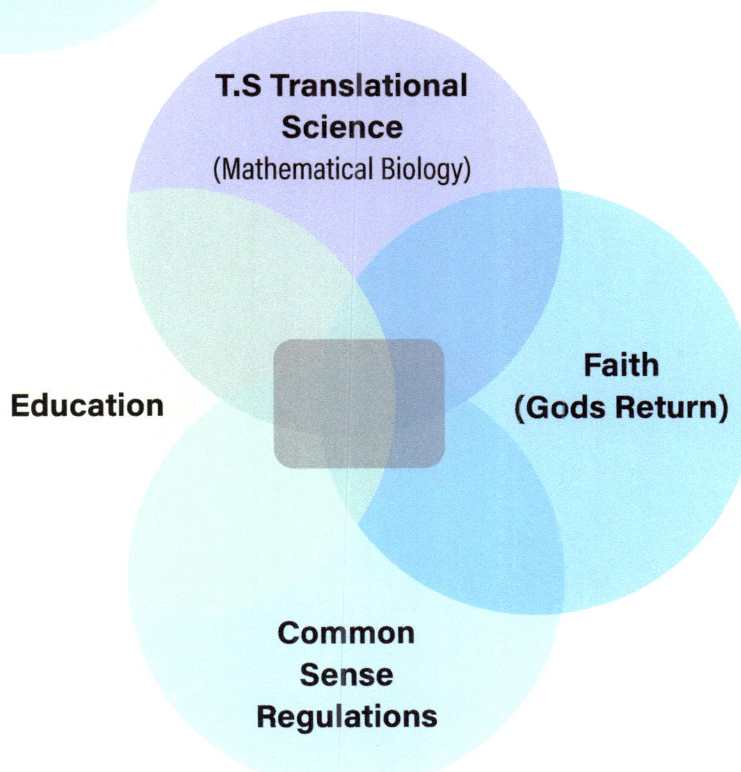
T.S Translational
Science
(Mathematical Biology)
Education
Faith
(Gods Return)
Common
Sense
Regulations

Endnotes

1 Samuel S. Epstein, *The Politics of Cancer, Revisited* (East Ridge, TN: East Ridge Press, 1998). Annual Fund letter 1998, *The Politics of Cancer, Revisited.*

2 Epstein, *The Politics of Cancer, Revisited.*

3 Extracted from *HemOnc Today*, May 2021.

4 Annals of Internal Medicine, Oct 2017. Conversation with T Quill, MD.

5 US National Library of Medicine, "Injection of Bromelain and Acetylcysteine in Combination Into Recurrent Mucinous Tumor or Pseudomyxoma Peritonei," 2019/2020, https://clinicaltrials.gov/ct2/show/NCT03976973.

6 Extracted from *The Oncologist*, Vol 26, Aug 2021.

7 Randolph's world, March 2010 (*Samuel Epstein, Politics of Cancer, Revisited*)

8 Roger Stupp et al., "Radiotherapy Plus Concomitant and Adjuvant Temozolomide for Glioblastoma." *New England Journal of Medicine* 352, no. 2 (March 2005): 987-96, https://pubmed.ncbi.nlm.nih.gov/15758009/

9 "Cytotoxic Drugs Can Increase Cancer Cell Resistence," *Oncology Times* 43, no. 17 (September 2021): 34, https://doi.org/0.1097/01.COT.0000791856.54210.79.

10 OncologyLive, Head and Neck Cancer, Dec 2021.

11 José Manuel Pérez-García et al., "Chemotherapy De-escalation Using an 18 F-FDG-PET-based Pathological Response-Adapted Strategy in Patients with HER-2-positive Early Breast Cancer (PHERGain)," *The Lancet* 22, no. 6 (2021): 858-71.

12 Michael Gnant et al., "Duration of Adjuvant Aromatase-Inhibitor Therapy in Postmenopausal Breast Cancer," *New England Journal of Medicine* 385 (July 2021): 395-405, https://www.nejm.org/doi/full/10.1056/NEJMoa2104162.

13 Deborah K. Mayer et al., "American Society of Clinical Oncology Clinical Expert Statement on Cancer Survivorship Care Planning," *Journal of Oncology Practice* 10, no. 6 (November 2014): 345-51, https://ascopubs.org/doi/10.1200/JOP.2014.001321.

14 John Marshall and Lisa Marshall, *Off Our Chests: A Candid Tour Through the World of Cancer* (IdeaPress Publishing, 2021).

15 Maurie Markman, review of *Off Our Chests* by John Marshall and Lisa Marshall, *Washington Post,* April 27, 2021, https://www.washingtonpost.com/entertainment/books/2021-books-may/2021/04/27/e200412e-a75a-11eb-8c1a-56f0cb4ff3b5_story.html.

16 L, Travis, et al, JCO, Nov 2021.

17 Nathan S. Gasek et al., "Strategies for Targeting Senescent Cells in Human Disease," *Nature Aging* 1 (2021): 870-79, https://www.nature.com/articles/s43587-021-00121-8.

18 Jeanne S. Mandelblatt et al., "Applying a Life Course Biological Age Framework to Improve the Care of Individuals With Adult Cancers," *JAMA Oncology* 7, no. 10 (November 2021), https://doi.org/10.1001/jamaoncol.2021.1160.

19 “Dasatinib Beneficial in Type 2 Diabetes,” Drugs.com, November 24, 2021, https://www.drugs.com/news/dasatinib-beneficial-type-2-diabetes-101882.html.

20 Giorgia Mangili et al., “Recurrent Granulosa Cells Tumors (GCTs) of the Ovary: A MITO-9 Retrospective Study,” *Gynecologic Oncology* 130, no. 1 (July 2013): 38-42, https://pubmed.ncbi.nlm.nih.gov/23623833/.

21 Matteo Lambertini et al., “Final Analysis of the PROMISE-GIM6 Phase III Trial Assessing GnRH Agonist Use During Chemotherapy to Preserve Ovarian Function in Premenopausal Patients with Early Breast Cancer,” Journal of Clinical Oncology 39, no. 15_suppl (May 2021): 516, https://ascopubs.org/doi/abs/10.1200/JCO.2021.39.15_suppl.516.

22 Charles Shapiro ASCO Post expert review 2021.

23 Zhouguang Hui et al., “Effect of Postoperative Radiotherapy for Patients With pIIIA-N2 Non-Small Cell Lung Cancer After Complete Resection and Adjuvant Chemotherapy,” *JAMA Oncology* 7, no. 8 (August 2021): 117-85, https://pubmed.ncbi.nlm.nih.gov/34165501.

24 Yukihide Kanemitsu et al., “Hepatectomy Followed by MFOLFOX6 Versus Hepatectomy Alone for Liver-Only Metastatic Colorectal Cancer (JCOG0603),” *Journal of Clinical Oncology* 39, no. 34 (2021): 3789-99. https://doi.org/10.1200/JCO.21.01032.

25 MK Frey 2014.

26 Massimo Di Maio et al., “Meta-Analysis of Single-agent Chemotherapy Compared with Combination Chemotherapy as Second-line Treatment of Advanced Non-small-cell Lung Cancer,” *Journal of Clinical Oncology* 27, no. 11 (2009): 1836-43.

27 Christopher Manz. "Palliative Care in Oncology." E. Chow et al., "How Accurate Are Physicians' Clinical Predictions of Survival and the Available Prognostic Tools in Estimating Survival Times In Terminally Ill Cancer Patients? *Clinical Oncology* 13, no. 3 (P209-18), https://www.clinicaloncologyonline.net/article/S0936-6555(01)99256-8/pdf.

28 Nina A Bickell et al., "Quality of Breast Cancer Care: Perception versus Practice," *Journal of Clinical Oncology* 30, no. 15 (2012): 1791-95.

29 Jennifer Southall. "Growing Focus on Patients' Mental Health Appears to Lower Suicidality, Improve Cancer Outcomes." *HemOnc*, April 23, 2021, https://www.healio.com/news/hematology-oncology/20210421/growing-focus-on-patients-mental-health-appears-to-lower-suicidality-improve-cancer-outcomes.

30 Jennifer Southall, "'Overdiagnosis' May Not Fully Explain Rise in Pediatric Thyroid Cancer Incidence," *HemOnc Today*, June 25, 2021, https://www.healio.com/news/hematology-oncology/20210623/overdiagnosis-may-not-fully-explain-rise-in-pediatric-thyroid-cancer-incidence.

31 Hila Shaim et al., "Targeting the αV Integrin/TGF-β Axis Improves Natural Killer Cell Function against Glioblastoma Stem Cells," *Journal of Clinical Investigation* 131, no. 14 (June 2021): e142116, https://www.jci.org/articles/view/142116.

32 Corinne A. Nief et al., "Resetting the Tumor Microenvironment to Favor Anti-tumor Immunity After Local Ablation," *Journal of Clinical Oncology* 39, no. 15_suppl (2021): 2561-61, https://ascopubs.org/doi/abs/10.1200/JCO.2021.39.15_suppl.2561.

33 Luigi Sapio et al., "The Natural cAMP Elevating Compound Forskolin in Cancer Therapy: Is It Time?" *Journal of Cellular Physiology* 232, no. 5 (2017): 922-27 https://pubmed.ncbi.nlm.nih.gov/27739063/; Michela Illiano et al., "Forskolin Sensitizes Human Acute Myeloid

Leukemia Cells to H3K27me2/3 Demethylases GSKJ4 Inhibitor via Protein Kinase A," *Frontiers in Pharmacology* online (July 2018), https://doi.org/10.3389/fphar.2018.00792.

34 Marina Gobbe Moschetta-Pinheiro et al., "35 Decreased Epithelial Mesenchymal Transition Process After AGTR-1 Gene Edition By Crispr/Cas9, Losartan, and PARP Inhibitor Treatment in Breast Cancer Cell Line," 38th Annual Miami Breast Cancer Conference Abstracts Suppl 1, https://www.cancernetwork.com/view/35-decreased-epithelial-mesenchymal-transition-process-after-agtr-1-gene-edition-by-crispr-cas9-losartan-and-parp-inhibitor-treatment-in-breast-cancer-cell-line.

35 N. Jewel Samadder et al., "Comparison of Universal Genetic Testing vs Guideline-Directed Testing for Patients With Hereditary Cancer Syndrome," *JAMA Oncology* 7, no. 2 (2021): 230-37, https://jamanetwork.com/journals/jamaoncology/fullarticle/2772576.

36 Siddhartha Devarakonda et al., "Genomic Profiling of Lung Adenocarcinoma in Never-Smokers," *Journal of Clinical Oncology* 39, no. 33 (November 2021): 3747-58.

37 Peter A. Fasching et al., "BRCA1/2 Mutations and Bevacizumab in the Neoadjuvant Treatment of Breast Cancer," Journal of Clinical Oncology 36, no. 22 (August 2018): 2281-87, https://pubmed.ncbi.nlm.nih.gov/29791287.

38 Zsofia K. Stadler et al., "Therapeutic Implications of Germline Testing in Patients With Advanced Cancers," *Journal of Clinical Oncology* 39, no. 24 (August 2021): 2698-709, https://ascopubs.org/doi/abs/10.1200/JCO.20.03661.

39 Anand R. Habib et al., "Recommendations From Breast Cancer Centers for Frequent Screening Mammography in Younger Women May Do More Harm Than Good," *JAMA Internal Medicine* 181, no. 5 (March 2021): 588-89, https://jamanetwork.com/journals/jamainternalmedicine/article-abstract/2777518.

40 Swedish study of HPV vaccination and risk of invasive cervical cancer, 2021 https://pubmed.ncbi.nlm.nih.gov/32997908/

41 Trevor J. Pugh et al., "The Genetic Landscape of High-risk Neuroblastoma," *Nature Genetics* 45, no. 3 (March 2013): 279-84, https://www.ncbi.nlm.nih.gov/pmc/articles/PMC3682833/.

42 Faren Nadeu et al. https://www.ncbi.nlm.nih.gov/pmc/articles/PMC8191863/

43 Miijin Kim, Hem Onc 2021 https://www.mdpi.com/2227-9059/9/11/1554/htm

44 clinical advances in hem/ Onc, 2021, Volume 19

45 Brain Cancer, Oncology Times, July 2021 https://journals.lww.com/oncology-times/fulltext/2021/07050/pediatric_treatment_options_for_average_risk.11.aspx

46 JAMA oncology 2021,7(9) 1313. https://jamanetwork.com/journals/jamaoncology/article-abstract/2782033

47 Sarfaty, Michal et al. 2021. Genitourinary Medical Oncology Expert Opinion Survey Regarding Treatment Management in the COVID-19 Pandemic. Clinical Genitourinary Cancer. 19(3) pp. e178-83

48 Oncology Times, Volume 43, Number 7, April 5, 2021, pp. 16-17

49 National Comprehensive Cancer Network. https://www.nccn.org/docs/default-source/covid-19/2021_covid-19_vaccination_guidance_v5-0.pdf?sfvrsn=b483da2b_74

50 U.S. National Library of Medicine. 2014/2022. Individualized Treatment in Treating Patients with Stage II-IVB

Nasopharyngeal Cancer Based on EBV DNA. https://clinicaltrials.gov/ct2/show/NCT02135042

51 Eliran Kadosh et al., "The Gut Biome Switches Mutant P53 from Tumour-suppressive to Oncogenic," *Nature* 586, no. 7827 (2020): 133-38, https://pubmed.ncbi.nlm.nih.gov/32728212.

52 Oncology Times, Volume 43, Number 7, April 5, 2021 https://www.nursingcenter.com/journalinfo?YY=&Journal_ID=401957&Issue_ID=5831215&About=1

53 Journal of National Comprehensive Cancer, Volume 19, Issue 8, Aug 2021.

54 American Association for Cancer Research Virtual Conference 2021

55 Developmental Therapeutics 2021

56 Ying Ru Zhao et al., "NELSON Lung Cancer Screening Study," *Cancer Imaging* 11, no. 1A (2011): S79-84, https://www.ncbi.nlm.nih.gov/pmc/articles/PMC3266562/.

57 Alice Goodman, "Enzalutamide Improves Overall Survival in Hormone-Sensitive Metastatic Prostate Cancer," *The ASCO Post,* June 25, 2019, https://ascopost.com/issues/june-25-2019/enzalutamide-improves-overall-survival-in-hormone-sensitive-metastatic-prostate-cancer/.

58 Mark C. Markowski et al., "A Multicohort Open-label Phase II Trial of Bipolar Androgen Therapy in Men with Metastatic Castration-resistant Prostate Cancer (RESTORE)," European Urology 79, no. 5 (May 2021): 692-99, https://www.ncbi.nlm.nih.gov/pmc/articles/PMC7775877/; "Bipolar Androgen Therapy," CDMRP, June 23, 2020, https://cdmrp.army.mil/pcrp/research_highlights/20denmeade_highlight.aspx.

59 Theodoros Karantanos, Paul G. Corn, and Timothy C. Thompson, "Prostate Cancer Progression After Androgen Deprivation Therapy," *Oncogene* 32, no. 49 (December 2013): 5501-11, https://www.ncbi.nlm.nih.gov/pmc/articles/PMC3908870/; Yufei Liu and Haowen Jiang, "Compositional Differences of Gut Microbiome in Matched Hormone-sensitive and Castration-resistant Prostate Cancer," *Translational Andrology and Urology* 9, no. 5 (October 2020): 1937-44, https://www.ncbi.nlm.nih.gov/pmc/articles/PMC7658119/.

60 Eric V. Li et al., "Efficacy and Adverse Effects of Docetaxel for Metastatic, Hormone-sensitive Prostate Cancer Among Elderly Men: A Post Hoc Analysis of the CHAARTED Trial," *Clinical Genitourinary Cancer* 19, no. 5 (2021): 388-95.

61 Erica Mayer. https://www.researchgate.net/scientific-contributions/Erica-L-Mayer-39389791

62 Michel Bolla et al., "Short Androgen Suppression and Radiation Dose Escalation in Prostate Cancer," *Journal of Clinical Oncology* 39, no. 27 (September 2021): 3022-33.

63 Caroline Helwick, "After Surgery for Colorectal Cancer, Intensive Monitoring of Little Benefit: PRODIGE 13," *The Asco Post,* November 10, 2020, https://ascopost.com/issues/november-10-2020/after-surgery-for-colorectal-cancer-intensive-monitoring-of-little-benefit-prodige-13/.

64 Umberto Basso et al., "Prognostic Role of Circulating Tumor Cells in Metastatic Renal Cell Carcinoma: A Large, Multicellular, Prospective Trial," *The Oncologist* 26, no. 9 (June 2021): 740-50.

65 Himalee, S. et al. JCO Dec 2021

66 Maria Schwaederle et al., "On the Road to Precision Cancer Medicine," *Molecular Cancer Therapy* 14, no. 6 (June 2015): 1488-94, https://pubmed.ncbi.nlm.nih.gov/25852059.

67 National Cancer Institute, "NCI-MATCH Trial (Molecular Analysis for Therapy Choice)," https://www.cancer.gov/about-cancer/treatment/clinical-trials/nci-supported/nci-match.

68 Zettler, Journal of Clinical Oncology, Nov 2021

69 Alex J. Liu et al., "Treatment Outcomes of Well-differentiated High-grade Neuroendocrine Tumors," The Oncologist, 26, no. 5 (May 2021): 383-88, https://pubmed.ncbi.nlm.nih.gov/33496040.

70 Satya Das, Taymayeh Al-Toubah, and Jonathan Strasberg, "Chemotherapy in Neuroendocrine Tumors," *Cancers (Basel)* 13, no. 19 (September 2021): 4872, https://pubmed.ncbi.nlm.nih.gov/34638356.

71 Sameer Agnihotri and Gelarah Zadeh, "Metabolic Reprogramming in Glioblastoma," Neurologic Oncology 18, no. 2 (February 2016): 160-72, https://www.ncbi.nlm.nih.gov/pmc/articles/PMC4724176/.

72 Sheilesh Advani et al, Clinical colorectal CA, June 2021

73 Wen Gao et al., "Selective Antitumor Activity of Ibrutinib in EGFR-mutant Non-small Cell Lung Cancer Cells," Journal of the National Cancer Institute 106, no. 9 (September 2014): dju204, https://pubmed.ncbi.nlm.nih.gov/25214559/.

74 Christine Roder and Melanie J. Thomson, "Auranofin: Repurposing an Old Drug for a Golden New Age," *Drugs in R&D* 15 (2015): 15-30, https://link.springer.com/article/10.1007/s40268-015-0083-y.

75 US National Library of Medicine, "Phase I and II Study of Auranofin in Chronic Lymphocytic Leukemia (CLL)," August 2011, https://clinicaltrials.gov/ct2/show/NCT01419691.

76 Chin-Yap Loh et al., "Signal Transducer and Activator of Transcription (STATs) Proteins in Cancer and Inflammation," Frontiers in Oncology 21 (February 2019), https://www.frontiersin.org/articles/10.3389/fonc.2019.00048/full.

77 Nan-Hoon Kim et al., "Antiproliferative Effect of Gold(I) Compound Auranofin Through Inhibition of STAT3 and Telomerase Activity in MDA-MB 231 Human Breast Cancer Cells," *BMB Reports* 46, no. 1 (January 2013), https://www.ncbi.nlm.nih.gov/pmc/articles/PMC4133824/; Jing Hu et al., "Auranofin Enhances Ibrutinib's Anticancer Activity in EGRF-mutant Lung Adenocarcinoma," *Molecular Cancer Therapeutics* 17, no. 10 (October 2018): 2156-64, https://mct.aacrjournals.org/content/molcanther/17/10/2156.full.pdf; US National Library of Medicine, "Sirolimus and Auranofin in Treating Patients with Advanced or Recurrent Non-small Cell Lung Cancer or Small Cell Lung Cancer," November 2012, https://clinicaltrials.gov/ct2/show/NCT01737502; Guo-Xin Hou et al., "Elimination of Stem-like Cancer Cell Side-population by Auranofin Through Modulation of ROS and Glycosis," Cell Death & Disease 9, no. 89 (2018), https://www.nature.com/articles/s41419-017-0159-4.

78 JNCCN, Vol 19, 2021

79 JNCCN, Vol 19, 2021.

80 Fox Chase Cancer Center study. Jennifer L. McQuade et al., "Association of Body-mass Index and Outcomes in Patients with Metastatic Melanoma Treated with Targeted Therapy, Immunotherapy, or Chemotherapy," *The Lancet* 19, no. 3 (March 2018): 310-22, https://www.thelancet.com/journals/lanonc/article/PIIS1470-2045(18)30078-0/fulltext?elsca1=tlxpr.

81 Rishi Jain et al., "The Role of Malnutrition and Muscle Wasting in Advanced Lung Cancer," Current Oncology Reports 22, no. 6 (May 2020): 54, https://pubmed.ncbi.nlm.nih.gov/32409907/.

82 Elizabeth Rose. Oncology Times, Volume 43, Issue 20, Oct 2021.

83 ASCO Vol 38 issue 22.

84 Kaysia Ludford et al., "Safety and Efficacy of Neoadjuvant Pembrolizumab in Mismatch Repair Deficient Localized/Locally Advanced Solid Tumors," *Journal of Clinical Oncology* 39, no. 15_suppl (May 2021):2520, https://ascopubs.org/doi/abs/10.1200/JCO.2021.39.15_suppl.2520.

85 Leonid Hanin, "Paradoxical Effects of Tumor Shrinkage on Long-term Survival of Cancer Patients," *Frontiers in Applied Mathematics and Statistics* (August 2020), https://www.frontiersin.org/articles/10.3389/fams.2020.00027/full.

86 Dibash Kumar. https://tlcr.amegroups.com/article/view/54551/html

87 Clinical Genito Urinary Cancer, Volume 19, Issue 3, June 2021.

88 Bray SM, LeeJ, Sci, rep, 2019. https://www.ncbi.nlm.nih.gov/pmc/articles/PMC3653905/

89 "Studies Suggest People with Blood Cancers May Not Be Optimally Protected after Covid-19 Vaccination" (press release), American Society of Hematology, April 16, 2021, https://www.hematology.org/newsroom/press-releases/2021/blood-herishanu-covid-vaccine-and-cll.

90 Amr Kamel Khalil Ahmed and Abdullah Alkattan, "Pitfalls at Chemistry of Adenoviral Vector Vaccine against Covid-19 and how to Circumvent It," ScienceOpen.com, June 27, 2021, https://www.scienceopen.com/hosted-document?doi=10.14293/S2199-1006.1.SOR-.PP96ICB.v1.

91 JAMA Oncology Nov 2021.

92 Nidal E. Muvarak et al., "Enhancing the Cytotoxic Effects of PARP Inhibitors with DNA Demethylating Agents—A Potential Therapy for Cancer," *Clinical Advances in Hematology & Oncology* 30, no. 4 (October 2021): 637-50, https://www.ncbi.nlm.nih.gov/pmc/articles/PMC5201166/

93 Singh et al JCO Oct 2021

94 Daniel F. Hayes, "Defining Clinical Utility of Tumor Biomarker Tests," *Journal of Clinical Oncology* 39, no. 3 (January 2021): 238-48, https://ascopubs.org/doi/full/10.1200/JCO.20.01572.

95 Heinz-Josef Lenz et al. https://www.researchgate.net/scientific-contributions/Heinz-Josef-Lenz-38617334

96 Amy Burd et al., "Precision Medicine Treatment in Acute Myeloid Leukemia Using Prospective Genome Profiling," *Nature Medicine* 26 (October 2020): 1852-58; https://www.nature.com/articles/s41591-020-1089-8; https://genomictestingcooperative.com/.

97 Mariano J. Alvarez et al., "Functional Characterization of Somatic Mutations in Cancer Using Network-based Inference of Protein Activity," *Nature Genetics* 48 (2016): 838-47, https://www.nature.com/articles/ng.3593.

98 Samia S. Messeha. https://www.frontiersin.org/articles/10.3389/fmolb.2018.00021/full

99 US National Library of Medicine, "Targeted Therapy Directed by Genetic Testing in Treating Patients With Advanced Refractory Solid Tumors, Lymphomas, or Multiple Myeloma (The MATCH Screening Trial)," February 3, 2022, https://clinicaltrials.gov/ct2/show/NCT02465060.

100 Marzia Del Re et al., "Understanding the Mechanisms of Resistance EGFR-positive NSCLC," *International Journal of Molecular Sciences* 20, no. 16 (August 2019): 3951, https://www.ncbi.nlm.nih.gov/pmc/articles/PMC6720634/.

101 Matthew Stenger, "Inhibition of SFRP2 Found to Inhibit Growth of Triple-negative Breast Tumors," *The ASCO Post,* May 15, 2012, https://ascopost.com/issues/may-15-2012/inhibition-of-sfrp2-inhibits-growth-of-triple-negative-breast-tumors/.

102 Edmund V. Capparelli et al., "Phase I Clinical Trial Results of Auranofin, a Novel Antiparasitic Agent," *Antimicrobial Agents and Chemotherapy* 61, no. 1 (January 2017): e01947-16, https://www.ncbi.nlm.nih.gov/pmc/articles/PMC5192119.

103 Arnold J. Levine and Moshe Oren, "The First 30 Years of p53: Growing Ever More Complex," Nature Reviews Cancer 9, no. 10 (October 2009): 749-58, https://pubmed.ncbi.nlm.nih.gov/19776744/.

104 Atushi Hoshino et al., "Inhibition of p53 preserves Parkin-mediated Mitophagy and Pancreatic Beta-cell Function in Diabetes," PNAS 111, no. 8 (2014), 3116-21, https://www.pnas.org/content/111/8/3116.

105 Elihu Estey JCO Vol 39 Issue 27

106 clinical advances in hemotology and oncology, Vol 19, issue 4

107 Olivia A Do et al.

108 Esra Akbay PhD Onc Times Sept 2021

109 Cristina Riccadonna et al., "Decitabine Treatment of Glioma-initiating Cells Enhances Immune Recognition and Killing," *PLoS ONE* 11, no. 8 (August 2016): e0162105, https://journals.plos.org/plosone/article?id=10.1371/journal.pone.0162105.

110 Claudia Bruedigam et al., "The Preclinical Efficacy of a Novel Telomerase Inhibitor, Imetelstat, in AML," Blood 128, no. 22 (2016): 578, https://ashpublications.org/blood/article/128/22/578/94846/The-Preclinical-Efficacy-of-a-Novel-Telomerase.

111 Stephen P. Hunger and Charles G. Mullighan, "Acute Lymphoblastic Leukemia in Children," The New England Journal of Medicine 373 (2015): 1541-52, https://www.nejm.org/doi/10.1056/NEJMra1400972.

112 Yang et al. https://www.sciencedirect.com/science/article/pii/S216225312100024X

113 Smith Clinical Advances in Hematology & Oncology Oct 2021

114 Monika Janda et al., "Complete Pathological Response Following Levonorgestral Intrauterine Device in Clinically Stage 1 Endometrial Adenocarcinoma," Gynecologic Oncology 161, no. 1 (April 2021): 143-151, https://www.gynecologiconcology-online.net/article/S0090-8258(21)00089-5/fulltext.

115 Chen et al. https://ascopubs.org/doi/full/10.1200/JCO.2012.42.7906

116 Robin K Kelly MD, New England Journal of Medicine, July 15, 2021

117 Shih-Han Kao et al., "Hypoxia, Epithelial-Mesenchymal Transition, and TET-mediated Epigenetic Changes," *Journal of Clinical Medicine* 5, no. 2 (February 2016): 24, https://www.ncbi.nlm.nih.gov/pmc/articles/PMC4773780.

118 https://ashpublications.org/bloodadvances/article/4/8/1670/454447/Targeted-next-generation-sequencing-of-circulating

119 Sze-Hwei Lee et al., "High Expression of Dedicator of Cytokinesis 1 (DOCK1) Confers Poor Prognosis in Acute Myeloid Leukemia," *Oncotarget* 8, no. 42 (September 2017): 72250-59, https://www.ncbi.nlm.nih.gov/pmc/articles/PMC5641127.

120 Gauthier Bouche, et al., Journal of Clinical Oncology

121 JCO December 2021

122 https://www.ncbi.nlm.nih.gov/pmc/articles/PMC8533995/

123 Y.K. Jin, X.H. Li, W. Wang, J. Liu, W. Zhang, Y.S. Fang, Z.F. Zhang, H.P. Dai, W. Ning, C. Wang. Follistatin-like 1 promotes bleomycin-induced pulmonary fibrosis through the transforming growth factor beta 1/mitogen-activated protein kinase signaling pathway. Chin. Med. J. (Engl.), 131 (2018), pp. 1917-1925

124 https://www.ncbi.nlm.nih.gov/pmc/articles/PMC7331721/

125 Kikuchi, Y., Kizawa, I., Oomori, K., Matsuda, M., and Kato, K. (1986). Adjuvant Effects of Tranexamic Acid to Chemotherapy in Ovarian Cancer Patients with Large Amount of Ascites. Acta Obstet. Gynecol. Scand. 65, 453–456. doi:10.3109/00016348609157383

126 Liu Yang et al., "Hexokinase 2 Discerns a Novel Circulating Tumor Cell Population Associated With Poor Prognosis in Lung Cancer Patients," PNAS 118, no. 11 (March 2021): e2012228118, https://www.pnas.org/content/118/11/e2012228118.

127 Gabriella Cirmena et al., "Circulating Tumor DNA Using Tagged Targeted Deep Sequencing to Assess Minimal Residual Disease in

Breast Cancer Patients Undergoing Neoadjuvant Chemotherapy," *Journal of Oncology* (January 2020), https://www.ncbi.nlm.nih.gov/pmc/articles/PMC7196957.

128 Paromita Raha et al., "Combined Histone Deacetylase Inhibition and Tamoxifen Induces Apoptosis in Tamoxifen-resistant Breast Cancer Models, by Reversing Bcl-2 Overexpression," *Breast Cancer Research* 17, no. 1 (February 2015): 26, https://pubmed.ncbi.nlm.nih.gov/25848915.

129 US National Library of Medicine, "Circulating Tumor DNA Testing in Predicting Treatment for Patients With Stage IIA Colon Cancer After Surgery," August 2019, https://clinicaltrials.gov/ct2/show/NCT04068103.

130 Xiaonan Tan et al., "A Curcumin-loaded Polymeric Micelle as a Carrier of a microRNA-21 Antisense-olgionucleotide for Enhanced Anti-tumor Effects in a Glioblastoma Animal Model," *Biomaterials Science* 6, no. 2 (January 2018): 407-17, https://pubmed.ncbi.nlm.nih.gov/29340361.

131 Journal of Oncology Sept 2021.

132 Khoa Pham et al., "Novel Glutamine Antagonist JHU395 Suppresses MYC-driven Medulloblastoma Growth and Induces Apoptosis," *Journal of Neuropathy & Experimental Neurology* 80, no. 4 (April 2021): 336-44, https://academic.oup.com/jnen/article-abstract/80/4/336/6169464.

133 Oncology Times: July 5, 2021 - Volume 43 - Issue 13 - p 22 doi: 10.1097/01.COT.0000765068.41879.4a

134 Heinz-Josef Lenz et al., "Impact of Consensus Molecular Subtype on Survival in Patients With Metastatic Colorectal Cancer," *Journal of Clinical Oncology* 37, no. 22 (August 2019): 1876-85, https://ascopubs.org/doi/10.1200/JCO.18.02258.

135 https://www.ncbi.nlm.nih.gov/pmc/articles/PMC8128439/

136 E.A. Klein et al., "Clinical Validation of a Targeted Methylation-based Multi-cancer Early Detection Test Using an Independent Validation Set," Annals of Oncology 32, no. 9 (September 2021): 1167-77, https://www.annalsofoncology.org/article/S0923-7534(21)02046-9/fulltext.

137 James Sun et al., "Impact of Axillary Dissection Among Patients With Sentinel Node–Positive Breast Cancer Undergoing Mastectomy," *Journal of the National Comprehensive Cancer Network* 19, no. 1 (2021): 40-47, https://pubmed.ncbi.nlm.nih.gov/33406495/.

138 Yasuhisa Hasegawa et al., "Neck Dissections Based on Sentinel Lymph Node Navigation Versus Elective Neck Dissections in Early Oral Cancers," *Journal of Clinical Oncology* 39, no. 18 (2019): 2025-36, https://ascopubs.org/doi/abs/10.1200/JCO.20.03637?af=R.

139 Oncology Times December 2021.

140 Laura M. Fernandez et al., "Conditional Recurrence-free Survival of Clinical Complete Responders Managed by Watch and Wait After Neoadjuvant Chemoradiotherapy for Rectal Cancer in the International Watch & Wait Database," *Lancet Oncology*, 22, no. 1 (January 2021): 43-50, https://pubmed.ncbi.nlm.nih.gov/33316218/.

141 Clinical Advances in Hematology & Oncology, Oct 2021.

142 Breast Cancer Prevention, Time for change, JCO oncology practice, Dec 2021.

143 Amy Body et al., "Stage III Colon Cancer: Is Neoadjuvant Chemotherapy Ready for Prime Time?" Digestive Medicine Research 4 (March 2021), https://dmr.amegroups.com/article/view/6951/html.

144 Femke CR stall, et al.

145 The Lancet Oncology January 2021 study in Japan.

146 Mitch Hayes, Michael Lin-Brande, and Sudhir Isharwal, "Primary Focal Therapy for Localized Prostate Cancer: A Review of the Literature," *Oncology (Williston Park)* 35, no. 5 (2021): 261-69, https://doi.org/10.46883/ONC.2021.3505.0261.

147 1 ASCO Daily News.

148 "Obesity Increases Survival in Advanced Prostate Cancer" (news release), Eurekalert, July 10, 2021, https://www.eurekalert.org/news-releases/846204#:~:text=The percent20study percent2C percent-20presented percent20today percent20at,thinner percent20patients percen-t20over percent2036 percent20months.

149 Erica Carpenter, Liquid Biopsy Laboratory.

150 Nisha Kurani and Emma Wager, "How Does the Quality of the U.S. Healthcare System Compare to Other Countries?" Peterson-KFF Health System Tracker, September 30, 2021, https://www.healthsystem-tracker.org/chart-collection/quality-u-s-healthcare-system-compare-coun-tries/#:~:text=Though percent20DALYs percent20have percent20declined percent20in,for percent20comparable percent20countries percent2C per-cent20on percent20average.

151 Selena Simmons-Duffin, "Poll Finds Public Health Has a Trust Problem," NPR, May 13, 2021, https://www.npr.org/2021/05/13/996331692/poll-finds-public-health-has-a-trust-problem.

152 The Oncologist vol 26 Sept 2021

INDEX

About the Author

Dr. Mohammad Amin Nezami is an MD graduate of USC and UCSF internships and residencies. He has been in practice for 23 years and currently practices in California.

He has 47 issued or pending patents in relation to cancer therapy and Covid-19 in the US and globally. His mission is to disrupt the current standard of care with a breakthrough that changes the way patients with cancer are treated, using nontoxic and yet effective therapies.

Dr. Nezami serves as a national and international speaker on oncology and epigenetic science. He has been involved in many research projects, publications, and presentations for organizations including the American Society of Clinical Oncology, the American Association for Cancer Research, the Molecular & Precision Medicine Tri-Conference, the World Molecular Imaging Society and World Pharmaceutical Congress, and the National Cancer Institute-American Association for Cancer Research Annual Meeting. He is also an author of oncology literature, with work published in *Journal of Clinical Oncology, Annals of Oncology, Cancer Science, Journal of Cancer Therapy, Journal of Precision Oncology*, and many others.

He co-authored Dr. Andrew Weil's *Textbook for (Integrative) Oncology* in 2014 and served as investigator for the Best Case Series Program by the National Institutes of Health and the National Cancer Institute.

He has also served as a speaker for Caris Life Sciences (Caris Oncology Laboratory) and has provided scientific expert review for several peer-reviewed oncology journals, including *Journal of Cancer Therapy* and *British Journal of Medicine and Medical Research*, as well as serving as a peer review expert for multidisciplinary scientific journals such as *Biology and Medicine*, *Science International*, *Journal of Advances in Medical and Pharmaceutical Sciences, Asian Journal of Medicine and Health, International Journal of Cancer and Tumor*, and *Archivum Immunologiae Et Therapiae Experimentalis*, among others.

His main areas of interest are customized oncology and epigenetics. He currently teaches and consults for other physicians and oncologists and is extensively involved in research (as a PI) and pursuing several clinical trials in treatment of advanced and refractory cancer.

CPSIA information can be obtained
at www.ICGtesting.com
Printed in the USA
BVHW061926021022
648363BV00003B/10